THE GREATEST ADVENTURE

BASIC RESEARCH THAT SHAPES OUR LIVES

THE

BASIC RESEARCH

GREATEST

THAT SHAPES

ADVENTURE

OUR LIVES

Eugene H. Kone and Helene J. Jordan, Editors

The Rockefeller University Press · New York 1974

Although this book has been made possible by a grant from the
National Science Foundation, any opinions, findings, conclusions, or
recommendations expressed herein are those of the authors and do
not reflect the views of the National Science Foundation.

Preface

THE GENESIS OF THIS VOLUME is due to the former director of the National Science Foundation, Dr. William D.
McElroy, who in 1971 invited the presidents of some of
the nation's leading scientific and engineering societies to
comment on the question: How has basic research led to
practical applications that have affected the lives of men?

The response was so enthusiastic, so full of riches of
principle and detail, that the National Science Foundation accepted a proposal from The Rockefeller University to prepare a volume that would help to underline and
dramatize the intimate relationship between the research
scientist in the laboratory and the consumer of often-
unexpected technological developments derived from
that research. Providing continuing support for this project have been two scientists: Dr. H. Guyford Stever,
present Director of the National Science Foundation,
and Dr. Frederick Seitz, President of The Rockefeller
University.

I believe it is important to state the intentions of the
book that has evolved from Dr. McElroy's initial idea, as
well as to indicate what it is *not* trying to do. It is an attempt to introduce a general audience to some of the
fruits of basic research that we enjoy today and that hold
great promise for mankind's future, as well.

The Greatest Adventure (the title is from Isaac
Asimov's eloquent introduction) is problem oriented,
and focuses on vital research in various scientific disciplines, ranging from molecular biology to physics, that
have special pertinence today. But the book was not
designed to be encyclopedic. Rather, it tries to provide a
road map through only a few areas of contemporary

knowledge. For instance, because of its complexities, medicine as such has been omitted deliberately, although several chapters touch on findings of dramatic potential for the well-being of mankind.

The eminent scientists who contributed the chapters have cooperated enthusiastically in bringing to the reader descriptions of past and present research directly related to our earth, life, health, and civilization, and they have demonstrated an uncommon level of dedication to the project. (For example, one manuscript came from mainland China, where an author was visiting, and a second came from a hospital, where it was completed despite the illness of its author.) An attempt has been made to retain the individuality of each author and, at the same time, to provide a coherence that demonstrates the relationship of each branch of science to every other branch and to the life of man.

We have benefited from the wisdom of our advisory editors—science writers who spent considerable time helping to find the direction the book should take—and from the counsel of our distinguished Scientific Advisory Committee, whose names appear elsewhere in the volume. Making our burden easier has been the good will and aid of the National Science Foundation and its staff members. We also are deeply indebted to Mr. William A. Bayless, Director of The Rockefeller University Press, who has given of his time, business counsel, and support, and to Mr. Reynard Biemiller, Assistant Director of the Press, whose expertise is responsible for the design of the book.

I must pay particular tribute to Helene Jordan, who has borne the burden of this project almost from its inception, and whose unflagging zeal and good judgment characterize every page.

EUGENE H. KONE
The Rockefeller University
July, 1973

ACKNOWLEDGMENTS

We are specially grateful to the distinguished members of our Scientific Advisory Committee, who read and commented on those manuscripts that deal with their specialties. They include: Drs. Jeremy Bernstein, Stevens Institute of Technology (physics); Maurice Ewing, the Marine Biomedical Institute, University of Texas (earth sciences); Kenneth L. Franklin, the American Museum - Hayden Planetarium (astronomy); Harold Gershinowitz, The Rockfeller University (environment); Bernard L. Horecker, Roche Institute of Molecular Biology (biological sciences); Neal E. Miller, The Rockefeller University (behavioral sciences); and William H. Stein, The Rockefeller University (chemistry). In addition, we wish to thank our Advisory Editors, all well-known science writers, whose help in launching the book was invaluable. They are Edward Edelson, Science Editor, New York *Daily News;* Daniel C. Gillmor; and Barry Richman, Editorial Director, New York Academy of Sciences. Suggestions also were offered by the Creative Science Program of New York University, including its director, Dr. Myron A. Coler, and the seminar members. That program was made possible, in part, by The Van Wezel Foundation.

Alfred Rosenthal, Public Affairs Officer of the National Science Foundation, and Rodney Nichols, Vice President of The Rockfeller University, have been most generous with their time and counsel.

Our thanks go also to Bern Dibner and Mrs. Adele Mattysse, Burndy Library, Norwalk, Conn.' Dr. Humberto Fernández-Morán, University of Chicago; Frank Forrester, Information Officer, U.S. Geological Survey; Frank Scott, Society of the Plastics Industry; F. Peter Simmons, Grumman Aerospace Corporation; Mrs. Andromache Sismenides, Maternal and Child Welfare Section, Agency for International Development; and Miss Hazel Whittaker, Educational Publications Officer, Publications and Reports Branch, Atomic Energy Commission. We are particularly indebted to Mrs. Fred Dodge for her organizational and secretarial expertise.

Contents

CIVILIZATION

Of What Use?

> *One may detest nature and despise science, but it becomes more and more difficult to ignore them. Science in the modern world is not an entertainment for some devotees. It is on its way to becoming everybody's business.*
> Theodosius Dobzhansky, in *The Biology of Ultimate Concern*

IT IS THE FATE of the scientist to face the constant demand that he show his learning to have some "practical use." Yet it may not be of any interest to him to have such a "practical use" exist; he may feel that the delight of learning, of understanding, of probing the universe, is its own reward. In that case, he might even allow himself the indulgence of contempt for anyone who asks more.

There is a famous story of a student who asked the Greek philosopher Plato, about 370 B.C., of what use were the elaborate and abstract theorems he was being taught. Plato at once ordered a slave to give the student a small coin that he might not think he had gained knowledge for nothing, then had him dismissed from the school.

The student need not have asked and Plato need not have scorned. Who would today doubt that mathematics has its uses? Mathematical theorems, which seem unbearably refined and remote from anything a sensible man can have any interest in, turn out to be absolutely necessary to such highly essential parts of our modern life as, for instance, the telephone network that knits the world together.

This story of Plato, famous for two thousand years, has not made matters plainer to most people. Unless the application of a new discovery is clear and present, most are dubious of its value.

A story about the English scientist Michael Faraday illustrates the point. In his time, he was an enormously popular lecturer, as well as a physicist and chemist of the first rank. In one of his lectures in the 1840s, he illustrated the peculiar behavior of a magnet in connection with a spiral coil of wire which was connected to a galvanometer that would record the presence of an electric current.

There was no current in the wire to begin with, but when the magnet was thrust into the hollow center of the spiral coil, the needle of the galvanometer moved to one side of the scale, showing that a current was flowing. When the magnet was withdrawn from the coil, the needle flipped in the other direction, showing that the current was now flowing the other way. When the magnet was held motionless in any position within the coil, there was no current at all, and the needle was motionless.

At the conclusion of the lecture, one member of the audience approached Faraday and said, "Mr. Faraday, the behavior of the magnet and the coil of wire was interesting, but of what possible use can it be?" Faraday answered politely, "Sir, of what use is a newborn baby?"

It was precisely the phenomenon whose use was questioned so peremptorily by one of the audience that Faraday made use of to develop the electric generator, which, for the first time, made it possible to produce electricity cheaply and in quantity. That, in turn, made it possible to build the electrified technology that surrounds us today and without which life, in the modern sense, is inconceivable. Faraday's demonstration was a new-born baby that grew into a giant.

Even the shrewdest of men cannot always judge what is useful and what is not. There never was a man so ingeniously practical in judging the useful as Thomas Alva Edison, surely the greatest inventor who ever lived, and we can take him as our example.

In 1868, he patented his first invention. It was a device to record votes mechanically. By using it, congressmen could press a button and all their votes would be recorded and totaled instantly. There was no question that the invention worked; it remained only to sell it. A congressman whom Edison consulted, however, told him, with mingled amusement and horror, that there wasn't a chance of the invention being accepted, however unfailingly it might work. A slow vote, it seemed, was sometimes a political necessity. Some congressmen might have their opinions changed in the course of a slow vote, whereas a quick vote might, in a moment of emotion, commit the Congress to something undesirable.

Edison, chagrined, learned his lesson. After that, he decided never to invent anything unless he was sure it would be needed and wanted and not merely because it worked.

He stuck to that. Before he died, he had obtained nearly 1,300 patents—300 of them over a four-year stretch, or one every five days, on the average. Always he was guided by his notion of the useful and the practical.

On October 21, 1879, he produced the first practical electric light, perhaps the most astonishing of all his inventions. (We need only sit by candlelight for a while during a power breakdown to discover how much we accept and take for granted the electric light.)

In succeeding years, Edison labored to improve the electric light and, mainly, to find ways of making the glowing filament last longer before breaking. As was usual with him, he tried everything he could think of. One of his hit-or-miss efforts was to seal a metal wire into the evacuated electric light bulb, near the filament but not touching it. The two were separated by a small gap of vacuum.

Edison then turned on the electric current to see if the

presence of a metal wire would somehow preserve the life of the glowing filament. It didn't, and he abandoned the approach. However, he could not help noticing that an electric current seem to flow from the filament to the wire across that vacuum gap.

Nothing in Edison's vast practical knowledge of electricity explained that phenomenon, and all Edison could do was to observe it, write it up in his notebooks, and, in 1884 (being Edison), patent it. The phenomenon was called the "Edison effect," and it was the inventor's only discovery in pure science. Edison could see no use for it. He therefore pursued the matter no further and let it go, while he continued the chase for what he considered the useful and practical.

In the 1880s and 1890s, however, scientists who pursued "useless" knowledge for its own sake, discovered that subatomic particles (eventually called "electrons") existed, and that electric current was accompanied by a flow of electrons. The Edison effect was the result of the ability of electrons, under certain conditions, to travel unimpeded through a vacuum.

In 1904, the English electrical engineer John Ambrose Fleming (who had worked in Edison's London office in the 1880s in connection with the developing electric-light industry) made use of the Edison effect and of the new understanding that the electron theory had brought. He devised an evacuated glass bulb with a filament and wire which would let current pass through in one direction but not in the other. The result was a "current rectifier."

In 1906, the American inventor Lee De Forest made a further elaboration of Fleming's device, introducing a metal plate that enabled it to amplify electric current as well as to rectify it. The result is called a "radio tube" by Americans.

It is called that because only such a device could handle an electric current with sufficient rapidity and delicacy to make the radio a practical instrument for receiving and transmitting sound carried by the fluctuating amplitude of radio waves. In fact, the radio tube made all of our modern electronic equipment possible—including television.

The Edison effect, then, which the practical Edison shrugged off as interesting but useless, turned out to have more astonishing results than any of his practical devices. In a power breakdown, candles and kerosene lamps can substitute (however poorly) for the electric light, but what substitute is there for a television screen? We can live without it (if we consider it only as an entertainment device, which does it wrong), but not many people seem to want to.

In fact, the problem isn't a matter of showing that pure science can be useful. It is a much more difficult problem to find some branch of science that *isn't* useful. Between 1900 and 1930, for instance, theoretical physics underwent a revolution. The theory of relativity and the development of quantum mechanics led to a new and more subtle understanding of the basic laws of the universe and of the behavior of the inner components of the atom.

None of it seemed to have the slightest use for mankind, and the scientists involved—a brilliant group of young men—apparently had found an ivory tower for themselves that nothing could disturb. Those who survived into later decades looked back on that happy time of abstraction and impracticality as a Garden of Eden out of which they had been evicted. For out of that abstract work there unexpectedly came the nuclear bomb, and a world that now lives in terror of a possible war that could destroy mankind in a day.

But it did not bring only terror. Out of that research also came radio-isotopes, which have made it possible to probe the workings of living tissue with a delicacy otherwise quite impossible, and whose findings have revolutionized medicine in a thousand ways. There are also nuclear power stations, which, at present and in the future, offer mankind the brightest hope of ample energy during all his future existence on earth.

There is nothing, it turns out, that is more practical,

more downright important to the average man, whether for good or for evil, than the ivory-tower researches of the young men of the early twentieth century who could see no use in what they were doing and were glad of it, for they wanted only to revel in knowledge for its own sake.

The point is that we cannot foresee the consequences in detail. Plato, in demonstrating the theorems of geometry, did not envisage a computerized society. Faraday knew that his magnet-induced electric current was a new–born baby, but he surely did not foresee our electrified technology. Edison certainly didn't foresee a television set when he puzzled over the electric current that leaped the vacuum, and Einstein, when he worked out the equation $e = mc^2$ from purely theoretical considerations in 1905, did not sense the mushroom cloud as he did so.

We can only make the general rule that, through all of history, an increased understanding of the universe, however out-of-the-way a particular bit of new knowledge may seem, however ethereal, however abstract, however useless, has always ended in some practical application (even if sometimes only indirectly).

The application cannot be predicted, but we can be sure that it will have both its beneficial and its uncomfortable aspects. (The discovery of the germ theory of disease by Louis Pasteur in the 1860s was the greatest single advance ever made in medicine and led to the saving of countless millions of lives. Who can quarrel with that? Yet it also has led, in great measure, to the dangerous population explosion of today.)

It remains for the wisdom of mankind to make the decisions by which advancing knowledge will be used well and not ill, but all the wisdom of mankind will never improve the material lot of man unless advancing knowledge presents it with the matters over which it can make those decisions. And when, despite the most careful decisions, there come dangerous side-effects of the new knowledge, only still-further advances in knowledge will offer hope for correction.

And now we stand in the closing decades of the twentieth century, with science advancing as never before in all sorts of odd, and sometimes apparently useless, ways. We've discovered quasars and pulsars in the distant heavens. Of what use are they to the average man? Astronauts have brought back rocks from the moon at great expense. So what? Scientists discover new compounds, develop new theories, work out new mathematical complexities. What for? What's in it for you?

No one knows what's in it for you right now, any more than Plato knew in his time, or Faraday knew, or Edison knew, or Einstein knew.

But *you* will know if you live long enough; and if not, your children or grandchildren will know. And they will smile at those who say, "But what is the use of sending rockets into space?" just as we now smile at the person who asked Faraday the use of his demonstration.

In fact, unless we continue with science and gather knowledge, whether or not it seems useful on the spot, we will be buried under our problems and find no way out. Today's science is tomorrow's solution—and tomorrow's problems, too—and, most of all, it is mankind's greatest adventure, now and forever.

EARTH

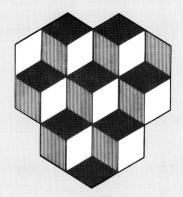

Any consideration of the world we live in today must start with its very beginnings, because today is part of yesterday and both are part of tomorrow. We could not have reached our present level of civilization without first having inquired into our origins by attempting to understand the laws of the natural world. Our first question, when we speak of such origins, must be: "Where did this earth of ours come from?" Then, in logical succession, we would ask about the water, the air, and the land from which our existence stems and which stand today in such peril. These subjects have gone far beyond the realm of idle curiosity; they present problems of such magnitude that the best minds in science today and those of tomorrow will be needed to find solutions. The four chapters that follow describe what scientific research has already discovered and what remains to be done.

CHAPTER 1 SIR FRED HOYLE

Cosmology and its Relation to the Earth

SOCRATES: *Shall we make astronomy the next study? What do you say?*
GLAUCON: *Certainly. A working knowledge of the seasons, months, and years is beneficial to everyone, to commanders as well as to farmers and sailors.*
SOCRATES: *You make me smile, Glaucon. You are so afraid that the public will accuse you of recommending unprofitable studies.*
Plato, *Republic* VII, circa 370 B.C.

FROM TIME IMMEMORIAL, man has looked up at the night sky and wondered about the pattern of the stars. It is scarcely possible to stand high on some remote hilltop, well away from city lights, gazing at the great arch of the Milky Way, without feeling a deep sense of purpose in the scheme of things. It is the business of the astronomer and the physicist to find out as much as they can about the relation of this outside world to our everyday experience. How has it led to our being where we are, here on earth?

Most astronomers now agree that earth and the other planets of our solar system were formed in the process that led to the condensation of the sun out of a cloud of gas. The gas was initially diffuse and spread out through the regions between the stars, like the clouds which astronomers observe today (Figure 1). Clouds of stars born perhaps as recently as a few million years ago also can be observed readily (Figure 2), so it makes good, logical sense to suppose that our sun was born in a similar way. When? Surprisingly enough, this question can be answered rather precisely—4.65 billion years ago. This is known from the careful investigation of certain radioactive substances that serve as precise fossil clocks for dating the age of earth, our moon, and many different meteorites. The same kind of technique has been used to date

FIGURE 1   *The Orion Nebula is a cloud of gas in which stars are being formed. Our sun and its planets are believed to have originated in a cloud of this kind.*

the whole Milky Way system. The result in this case is not quite so precise. The best answer lies somewhere between 12 and 15 billion years. Evidently our solar system—our sun and planets—have existed for only about 30 per cent of the age of our galaxy (i.e., the Milky Way system).

Astronomers are in much less agreement about the way in which the planets were formed, but as more and more information comes along, the area of disagreement becomes less. Solving the origin of the planets is rather like a complicated detective story, in which significant clues must be separated from a host of lesser ones. Perhaps the three most significant clues are the following.

1. The sun spins very slowly.
2. The planets divide into two groups—Mercury, Venus, Earth, and Mars on the inside—Jupiter, Saturn, Uranus, and Neptune on the outside. (Pluto may well be only an escaped satellite of Neptune, and is not significant here.)

FIGURE 2   *The Pleiades, a cluster of recently formed stars.*

3. The outer planets are composed almost wholly of volatile materials that are very abundant in the sun, whereas the inner planets are composed of much rarer refractory materials, such as rock and metal.

The problem is to weave a picture of the origin of the planets from these clues.

Observations have shown that the gas seen in Figure 1 possesses complex, internal, swirling motions. As a result, any average portion of the gas that condenses into a star must develop a spin. Moreover, when a mass of gas condenses, the rate at which it spins becomes faster as the system becomes more compact. This spin-up is demanded by the basic laws of mechanics—the principle of the conservation of angular momentum. It must take place inevitably, unless the spin becomes subject in some way to an external braking agency. Without such a braking agency, as calculation for the gas cloud of Figure 1 has shown, the rate of spin of a condensed star like the sun would be much faster than it actually is. So we can conclude that a braking process must have been operative when the sun was forming, and the first immediate step in our story is to identify that process.

The solution is simply this. The planetary material was shed by the condensing sun, which at first was spinning quickly. This led to an outer ring of planetary material that surrounded the sun. Then an action-reaction was set up between the planetary material and the sun, slowing the sun's spin and, at the same time, forcing the planetary material further and further outward—again because of the principle of conservation of angular momentum. The planetary material cooled as it went, so that various substances began to condense within it, initially as smoke and as droplets. Those substances that resist evaporation—rock and such common metals as iron—were the first to condense, and out of these materials were formed the innermost planets, including the earth. As the gases continued their outward journey, the temperature continued to lower and less refractory materials, such as water, began to condense. These formed the outer planets. In this way, we have tied our three clues together into a consistent picture.

Initially, the earth was hot, probably about the temperature of a blast furnace. Because of the high temperature, water and other comparatively volatile substances could not condense near the earth and must have been carried to the outer part of the solar system, where they helped to form the big planets—Jupiter, Saturn, Uranus, and Neptune. Hence, it seems unlikely that the life-forming volatiles—water and compounds of carbon—could have been present on earth in the first place. It seems they arrived here in a less direct—and in a rather curious—way.

During the detailed formation of the big planets, there must have been a great deal of dynamic activity. Vast swarms of smaller bodies must have been formed first. These bodies must have collided frequently and, as a result, an enormous spray must have developed everywhere throughout our solar system. Probably the comets are surviving pieces from that spray. The pieces consisted of icy bodies—some of ordinary ice, some of dry ice, others of more esoteric substances, such as hydrocyanic acid (which is thought by many chemists to have played a key role in the origin of life). They flew like great snowballs throughout the solar system, and many of them must have hit the newly formed earth. Besides scarring the surface, as the moon has been scarred by such impacts, these collisions added water and other life-forming substances to the earth. It is possible, perhaps likely, that we humans owe our existence to that early rain of vast snowballs.

Nothing in all these speculations is particularly special to our sun or to our planetary system. Stars otherwise similar to the sun spin slowly, too, so we might expect that a very large number of stars with planetary systems exist in the Milky Way. Nor is there anything in the chemistry of the process of planetary formation that seems special. Other stars are composed of the same materials as the sun, so very probably their planets, too, are like ours. That is, the refractory materials, mostly rock and metal, are always on the inside and the volatile substances—water, ammonia, methane—are always on the outside. As far as astronomy is concerned, other systems are probably very similar to our own—perhaps ten thousand million of them in our own galaxy.

Biochemical research going on in many parts of the world is coming closer to understanding the origin of terrestrial life. For instance, most chemists and biologists would agree that the primitive components of life are water, ammonia, methane, and hydrocyanic acid. These are the same building blocks that astronomers are finding to exist in large quantities in the gas clouds of the Milky Way. We still aren't sure how these basic building blocks came together into the exceedingly complex forms that are found in living cells. Yet, as vague speculation becomes replaced more and more by precise knowledge, nothing exclusive to our own solar system has been found (Table I).

Very recently, amino acids—without which life could not exist—have been found imbedded in meteorites. That exciting find has presented the possibility that these components may have been in the solar system when the earth first was formed. And that speculation, in turn, means that it is not impossible for billions of life systems to exist in various parts of our galaxy—and in other galaxies, too. In other words, astronomy and cosmology have brought us nearer to the conclusion that evolution from lower to higher forms of life is an inevitable result of what has been called the "uniformity of nature." That is, stars in our galaxy are very much alike in chemical composition; other galaxies, too, apparently resemble ours.

This brings up a question that man has asked himself for thousands of years. If life is indeed widespread, could we send signals out into space to be received and decoded by creatures on some planet moving around some distant star? Not quite. Our technology is not adequate for such a feat—yet. But if we are able to make as tremendous strides in electronic technology during the next fifty years as we have during the past fifty, the com-

TABLE I   *Molecules Found in the Interstellar Medium*

Discovery	Formula	Molecule
1937	CH	
1940	CN	Cyanogen
1941	CH +	
1963	OH	Hydroxyl
1968	NH_3	Ammonia
1968	H_2O	Water
1969	H_2CO	Formaldehyde
1970	CO	Carbon monoxide
1970	CN	Cyanogen
1970	H_2	Hydrogen
1970	HCN	Hydrogen cyanide
1970	?	X-ogen
1970	HC_3N	Cyano-acetylene
1970	CH_3OH	Methyl alcohol
1970	CHOOH	Formic acid
1971	CS	Carbon monosulphide
1971	NH_2CHO	Formamide
1971	SiO	Silicon oxide
1971	OCS	Carbonyl sulphide
1971	CH_3CN	Acetonitrile
1971	HNCO	Isocyanic acid
1971	HNC	Hydrogen isocyanide
1971	CH_3C_2H	Methyl-acetylene
1971	CH_3CHO	Acetaldehyde
1971	H_2CS	Thioformaldehyde
1972	CH_2NH	Formaldimine
1972	H_2S	Hydrogen sulphide

munication dream could become a reality.

However, there is a basic snag in such a communication system, one that cannot be overcome. Even if we could "call" a distant planet, we'd have a long wait for an answer—a single interchange of messages would take hundreds of years. But then, all big changes in human society take many hundreds of years. So, while an interchange system would hardly be suitable for swapping recipes or gossip, it could have a tremendous impact on the slow evolution of civilization as a whole for one important reason. Our civilization seems to be carried along by deep-seated currents whose nature we understand only partially. If we could know where these currents are leading us—information we might expect to learn from creatures on other planetary systems who already have experienced much the same social difficulties as ours—the path to the future would be much less rocky and mysterious.

At a shallower level are many problems of detail for which it would be an advantage to know the answer ahead of time. For example, the decision made many years ago to permit unrestricted use of the internal combustion engine has led to unpleasant social consequences that certainly were not anticipated in the age of the model-T Ford. Like the pilot of a boat in difficult waters, we would like to know what shoals lie ahead of us. Possibly a galactic communication system will one day make that possible.

It is interesting how quickly one can pass from problems concerning the sun and solar system to issues that affect the universe in the large—to cosmology, in fact. In considering how planetary material separated from the condensing sun, we also considered an action-reaction that braked down the sun's spin and forced the planetary material farther outward in the solar system. What was the agency of these phenomena? It turns out that the processes are curiously similar to the one Kepler suggested for gravitation. Kepler thought the planets were kept moving in their present-day orbits by magnetic forces from the sun. As far as gravitation is concerned, this idea was quite wrong, but, remarkably enough, it is exactly what is needed to understand the early outward motion of the planetary material and the slowing of the sun's spin. Early in the present century, astronomers discovered that our present-day sun actually possesses a magnetic field. Thus, it is entirely reasonable to suppose that the primeval sun also possessed a magnetic field, particularly as the fields that are found in sunspots would be amply strong enough to operate a braking process of the required kind.

This issue of the magnetic field is in itself a detail, albeit an important one. The issue becomes major, however, when we ask why the sun has a magnetic field in the first place. This question leads step by step to a consideration of the origin of the universe and to a difference between the "big-bang" and steady-state theories. We begin the sequence of steps by noticing that the question has never been answered in the sense of explaining an origin *in situ* and, indeed, there are strong grounds for thinking that the magnetic field of the sun did not originate spontaneously. Rather, we must say that the sun possesses a magnetic field because the gas cloud from which the sun condensed some four and a half billion years ago possessed a magnetic field. In other words, the magnetic field of the present-day sun is a fossil relic.

Very well, now let us go back another step and ask why the gas cloud from which the sun condensed possessed a magnetic field. Once again, it has been impossible to understand how such a field could arise spontaneously. We have to say that the localized gas cloud possessed a magnetic field because our galaxy as a whole possesses a magnetic field. And we have to say that our galaxy possesses a magnetic field because magnetic fields exist everywhere throughout the universe.

This brings us close to issues of cosmology, which is a science (or philosophy) that deals with the character of the universe as an orderly, natural system. We proceed by noticing that at no stage in the discussion that took us from our local solar system to the universe can we find a convincing argument for the origin of the field. At each stage we simply push the question back to the next stage. What happens now that we have reached the universe as a whole? Is there a further stage still to which we can appeal? Yes, indeed. We can appeal to an origin of the universe itself, thereby making the origin of the magnetic field an issue of cosmology.

We say that the universe possesses a magnetic field because that was the way the universe happened to be set up in the first place, or at least we reach this conclusion if we elect to follow the philosophy of big-bang cosmology. This curious and rather unsatisfactory situation applies to many other issues in the big-bang theory. Indeed, many aspects of astronomy and physics, some even affecting our everyday lives, can be explained only by the big-bang theory in terms of the way the universe was set up in the beginning.

The steady-state theory differs crucially from the big-bang theory in that it does not propose an origin of the whole universe, although even in the steady-state theory it is necessary to discuss the origin of limited parts of that universe. Let us deal first with the big-bang cosmology and its implications, because most astronomers at present favor it over the steady-state theory.

If, indeed, we should discover that the basic idea of the big-bang—namely, an origin for the whole universe—is indeed correct, undoubtedly we will gain a better understanding of what the concept of "origin" really means. However, instead of thinking in terms of an origin of the universe, it is probably better to conceive of some discontinuous transition, as if our universe—or at any rate an appreciable fraction of it—underwent a catastrophic change twenty to thirty billion years ago. It may well have done so from some preceding state. The precise nature of this catastrophic transition is likely to form the major topic of cosmology for many years to come. Our prospect of grappling successfully with a problem of this great depth would be much improved if we could find similar discontinuous transitions still occurring in the present-day universe. It seems possible that this may be so, although perhaps not on the scale of twenty to thirty billion years ago.

Fortunately, if major catastrophic events take place in the universe today, they can be studied directly with the aid of the wide range of techniques now available to the astronomer. In addition to optical and radio astronomy, we now have the newer disciplines of infrared and X-ray astronomy. Just as sound waves can have many frequencies of "pitches," so electric waves have frequencies that range from radio waves at the lower end to gamma rays at the upper end. The increasing frequencies progress from radio through millimeter waves, to infrared, to light, to ultraviolet, to X-rays, and finally to gamma rays. Each of these is essentially the same phenomenon— electromagnetic waves—but each has a different frequency. Radio waves and light waves can come through the earth's atmosphere with comparative freedom. Millimeter waves and some ranges of the infrared can come through only partially. The astronomer who uses ground-based equipment must, of necessity, restrict himself to these available sections of the electromagnetic "spectrum." Only by using rockets or satellites that go above the atmosphere can there be access to certain ranges of the millimeter band and of the infrared band, and all of the high-frequency part of the spectrum— ultraviolet, X-rays, and gamma rays.

Our technology has already progressed to this point, as is described by Dr. Schawlow in this book. Equipment carried in sounding rockets and satellites must be very small, compared to the large telescopes that can be erected on the ground. Consequently, experiments above the earth's atmosphere are, in general, rather simple in scope, compared to ground-based work. Even so, simple experiments in the new wavebands, especially X-rays and parts of the infrared band, have proved to be of great interest, because work at the frequencies in question cannot be done at all from the ground. Recent observations, both from the ground and from space, are bringing weight to bear on a whole range of significant problems. New electronic devices have also had an astronomical impact. These are currently in the process of transforming the older astronomical disciplines dramatically. Many types of telescopes and their ancillary instruments are now being controlled electronically, which permits observations to be made much more quickly, as well as more sensitively. This is discussed more fully in Chapter 11.

The now-rapid accumulation of new data is making it more and more likely that catastrophic events of a type that hitherto have been associated only with the origin of the whole universe actually happen locally on the scale of individual galaxies. It is interesting to ask whether a sufficiently large number of these local events might take the place of a single, major discontinuity of the whole universe. But perhaps there was no such discontinuity, no universal origin. This is the point of view of those who hold with steady-state cosmology.

Six or seven years ago, the evidence against the theory was thought to be strong, but it has weakened markedly in the past two years. More accurate and recent observations have shown that several apparent discrepancies between the predictions of the steady-state theory and older observations were simply the results of inaccuracies in the old observations. The only remaining issue—and it is an awkward one—for the steady-state theory is to explain the origin of a smooth background of radio waves, which observations show are found with remarkable uniformity all over the sky. This is the so-called microwave background.

At first thought, one might seek to attribute this radiation to a profusion of very distant, discrete sources of radio waves, similar to the many nearer sources that are actually known to exist. Most of these known sources are galaxies whose central regions have, for reasons that are still largely obscure, experienced enormous explosions. Calculation shows this idea to be inadequate, however. There simply are not enough sources of known types to explain the microwave background, either by the steady-state or the big-bang theory. In the big-bang theory, one can always appeal to the beginning of the universe, arguing that the microwave background is a survival, a fossil relic from the mode of origin of the universe. In the steady-state theory, on the other hand, no such recourse is possible, because it holds no such origin.

A new idea is needed, therefore, if the steady-state theory is to explain the observed microwave background. If a satisfactory new idea should be forthcoming eventually, the steady-state theory would certainly become as valid as the big-bang hypothesis. But a suitable new idea still remains to be found.

The fragmentation of science is a source of difficulty to all teachers and to all students—the connection of one research area to another is not always apparent. This is because science is rather like a vast and subtle jig-saw puzzle, and the usual way to attack a jig-saw puzzle is to work simultaneously on several parts of it. Only at the end do we seek to fit the different parts together into a coherent whole.

In science, one usually studies astronomy, geology, chemistry, and biology as separate disciplines. Yet we have seen here how all these subjects must merge if we are to attack in a meaningful way the questions that surround the origin and history of the earth, of the other planets of our solar system, and even of the universe itself. This synthesis of many aspects of science gives a clear indication of how future developments will move toward an increased understanding of how all aspects of science, from the components of the atom to the components of the universe, must ultimately be fitted together into a single pattern. The different parts of the jigsaw puzzle must come together at last.

CHAPTER 2 WALTER LANGBEIN

Water on the Land

*...we do not conclude that man cannot control floods
because he cannot make rain fall upward.*
George Gaylord Simpson, in *The Meaning of
Evolution*

IN THE PREVIOUS CHAPTER, we saw how the volatile components of our planet—water, ammonia, and methane—came to be formed. Man's dependency on the first of these is every day becoming more apparent and more pressing, although no mineral substance on the surface of the earth is more common. Water is pervasive—oceans, lakes, rivers, glaciers, swamps, soil water, ground water. The desert traveller of old who perished of thirst only a few yards above cool, fresh, ground water, dramatizes, in a way, man's constant quest for this essential for life. Surely water is not scarce, but one must know where to find it and have the means to get it.

Great civilizations of the past, now vanished from the earth, may have ended because of lack of water or proper water management, according to one theory. It has been suggested that the once-sophisticated cultures of the Fertile Crescent, which stretched from Palestine to Arabia, fell because of water shortages or excess salinity. The same fate apparently befell Mohenjo-Daro in India, a civilization that prospered more than three centuries B.C. The monsoon, which brought rain to the area, may have moved more to the east, causing drought and the eventual downfall of the realm. Our own Hohokum Indians in the Southwest also may have dispersed because of lack of water.

We all take water for granted until we feel its lack or its overabundance. Devastating floods, protracted droughts, and pollution demonstrate the broad gulf between the results of scientific research and public policy. This generation and those to come must have some understanding of what has been discovered about water and its occurrence if this vital resource is to be managed more wisely.

Despite its ubiquity, water is unique among natural substances. It is a mineral that is a renewable resource, circulating endlessly in an earthwide hydrological cycle; it is a realm among the earth sciences that is squarely relevant to nearly the whole gamut of current environmental anxieties and economic growth. What, precisely, are the properties of this remarkable substance?

Ordinary water is composed of two atoms of hydrogen and one of oxygen—H_2O. The water molecule is lopsided; the hydrogen atoms stick out and tend to approach the oxygen atoms of its neighboring water molecules. This is called "hydrogen bonding," and it explains many of the unique physical properties of water. For instance, the hydrogen bonds give water its strong cohesiveness and tensile strength (not unlike that of steel), and its adhesive property, which means that its oxygen atoms stick to, or wet, stones, flesh, and many other materials and so attract the hydrogen atoms. (For instance, oxygen atoms wet cotton but not the synthetic material nylon, which therefore is said to "drip dry.")

Water is truly a substance for all purposes. It is about the only abundant material that can exist in solid, liquid, and gaseous states—within the range of temperatures on earth's surface. Among solids, it is a lightweight and floats in its own liquid as ice. Its specific heat, which exceeds considerably that of other natural materials—for instance, it is five times greater than soil or stone—explains why the vast oceans as reservoirs of heat avoid thermal extremes and so make the earth habitable and why water is in such demand as a coolant. A supreme solvent, water carries the nutrients for life, and it is a medium for chemical reactions, a source of power, and a method of transporting waste.

THE HYDROLOGICAL CYCLE

Earth's stock of water is constant, endlessly recirculated, purified, and reused; it is never depleted or exhausted. This is known as the hydrological cycle (Figure 1). The cycle may be thought of as beginning and ending with the vast oceans, which, as they evaporate, feed vapor to the atmosphere and, in return, serve as sinks for the flow of rivers that drain the continents. The atmospheric vapor returns to earth as rain and snow. The precipitation that falls on the continents returns to the atmosphere as evaporation and as transpiration from life processes, such as that from the surfaces of green plants. The rest seeps into the ground or falls into lakes and rivers, so returning again to the sea. For instance, some 30 inches of rain and snow fall on continental United States each year, of which 21 inches returns to the atmosphere. The remaining nine inches, whether pumped from the ground or withdrawn from streams, is available to man.

Some idea of the quantities involved in this endless process of distillation, condensation, and drainage is given in Table I, which also shows the "residence time" of water in the several phases of the hydrological cycle. This time period varies from 4,000 years in the oceans to mere days in the atmosphere and rivers. The oceans are the storage basins; the atmosphere and rivers the conduits. So, although the concept of drinking once-used water is repugnant, one may gain consolation from the fact of replenishment. Recycling is only man's adaptation of nature's techniques.

As water moves through the hydrological cycle, it carries mineral salts in solution, chiefly sodium, calcium, chloride, and sulfate, which come from the slow erosion of rocks on all the continents and from salts derived from the oceans and recycled back to the land by fog, rain, snow, or mist. Some two billion tons of salt are carried

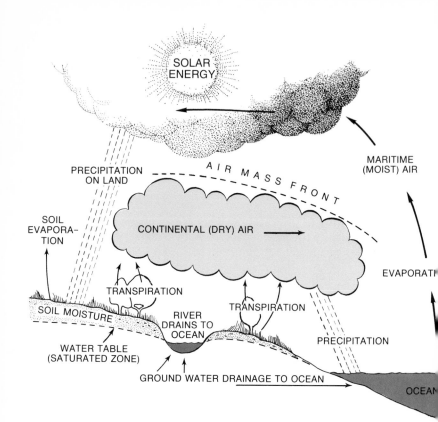

FIGURE 1 *The hydrological cycle. The earth's stock of water circulates endlessly; it is never exhausted.*

from the land to the sea each year, and man is contributing an increasing amount, largely via industrial waste. Combustion, life processes, and waste disposal ultimately reduce all of man's goods to mineral ash or volatile gas, which, whether on land or in the atmosphere, become entrained in the waters of the hydrological cycle. And as production of man's waste grows, recycling must contend with the ultimate disposal of mineral ash.

WATER TECHNOLOGY

The implements of water management are canals, dams, pipelines, wells, and pumps. In its main outlines, this

TABLE I. *The Hydrological Cycle*

Location	Volume (cubic miles)	Annual rate of circulation (cubic miles per year)	Residence time (years)
Oceans	320,000,000	80,000	4,000
Atmosphere	3,000	90,000	0.03
Ice caps and glaciers	7,000,000	1,000	7,000
Lakes	50,000	500	100
Rivers	300	5,000	0.06
Soil moisture	16,000	40,000	0.4
Shallow ground water	1,000,000	2,000	500
Deep ground water	1,000,000	<100	>10,000

basic technology would have been recognizable to Frontinus, the Augustinian water commissioner who described, in the first century A.D., the sources, aqueducts, and services of the city of Rome. Compared with other areas of modern industry today, water technology seems quite *un*-modern. Ideas evolve slowly as the scope of water planning enlarges from single-purpose, single (once-through) use, and single technology. A dam for water power alone is an example of a single-purpose, single-use, single-technology project. A dam built for water power plus irrigation is said to be multipurpose. "Multimethod" adds to multipurpose by including the combined use of river and ground water, recalmation and reuse of water, and desalination.

Although the principles are well known and citizens often become petulant about their neglect, the real problems of water management are economic. Nearly all river-flow is involved in some way with water use or control, and is not only considerably "cheaper than dirt"; it is the cheapest thing we can buy at 15 cents a ton delivered to one's home (with high quality-control), or about two cents a ton for untreated water delivered to an irrigation farm. These prices are so low that they do not attract new technology, so the "hardware" and machinery costs of water management also remain low. When, as with desalination of sea water, machinery and energy input increase, costs multiply by an order of magnitude.

Nevertheless, as water becomes more difficult to obtain, and as waste-disposal standards come under stringent control, one may anticipate increased use of the multitechnologies. Examples are the use of treated municipal waste water for industry, and recycling or closed-water systems that duplicate, in miniature, the natural hydrological cycle.

Another multimethod scheme might be a vast continental plumbing system to collect and distribute water of different qualities for specialized uses. Or one might conceive of using large subsurface galleries to collect, store, and treat, for its own use the rain that falls on a city, thus converting storm drainage from a waste to a resource. (A city with 40 inches of precipitation, 90 per cent of which can be recovered, receives enough water to supply 20,000 persons for each square mile on a once-through basis.) There is something hydrologically and intellectually satisfying about a community that is self-contained, as it were, with respect to its own water use and disposal.

WATER SUPPLY AND WATER USES

The steady, upward trend in the withdrawal of water from rivers, lakes, and the ground (Figure 2) against a resource which, unlike fossil fuels, does not change, has caused widespread anxieties that one day we will "run out of water." To be sure, there have been critical water shortages, especially during drought. Moreover, there are significant regional variations, based on the nature of the land. For example, the American Southwest experiences a state of perpetual shortage, to which the region re-

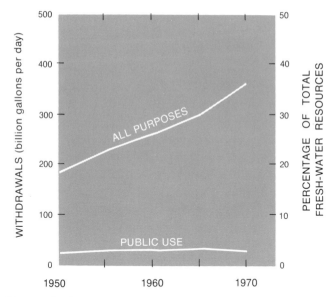

FIGURE 2 *Total water withdrawals in the United States from 1950 to 1970. (After the U.S. Geological Survey)*

sponds by long-distance transport and mining of ground-water resources where available, particularly in Arizona and the high plains of Texas. Eastern cities within the humid, well-watered regions of the country can experience severe shortages during prolonged drought (for example, that of the 1960s in the Northeast), when planning lags far behind the increase in use. Yet, alarmist views of impending natural shortages of water are not warranted. Supplies *can* be found, albeit at the price of new development and careful use.

As we have seen, the supply base of water molecules is unchanging. Even pollution has not yet reduced the base significantly, because water for public use can be and is treated to meet health and other quality standards. Present treatment practices, however, remove only those contaminants that are settleable, filterable, oxydizable, or killable. This covers a wide range of potential pollutants, but substances in solution are not removed by treatment. Irrigation, for example, increases the concen-

29 *Water on the Land*

tration of salts in solution. As a result, such water is impaired for subsequent use for other purposes. Toxic substances in solution have similar destructive results. Probably the most publicized of these—and the list is long—are pesticides and those attributed to the elements, such as radioactive isotopes. Less "bad" are those that are poisonous by nature of their composition—certain acids, for example—because they *will* decompose, albeit slowly. But complete purification methods *are* available and include distillation, which is a technique used in desalinating salt water. Distillation can make any stock of water molecules re-usable, but it is an expensive method. So this man-made problem can be solved only by men—those involved in water economics and water science, and by the consumer, who must be prepared to pay more for this commodity without which there would be no life.

Water can be hostile, as recurrent floods remind us. Every year, floods in the United States cause damage that averages several hundred million dollars. These range from the innumerable floods on the two- to three-million miles of small streams that dissect the national landscape to devastation on a vast scale that affects large streams and coastal areas. Often they are the result of man's ill–advised choices of places to build his homes, his factories, and his roads.

Although it has been dry land throughout most of its history, a flood plain is built by, and is an integral part of, the river system. Simply because it is flat, the flood plain is inviting, especially in hilly or mountainous territory where it may offer the sole level space for building or for highway and railway routes. Typically, a river conveys its ordinary flows within a well-defined channel bordering its flood plain, which it overflows about once every two to three years. At sporadic intervals of, say, 25, 50, or 100 years, a river may inundate its entire plain to a considerable depth (Figure 3).

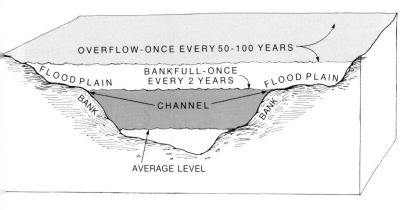

FIGURE 3 *Once every 50 to 100 years, a river overflows its flood plain to a depth that is about equal to the height of its banks.*

After the devastating floods in 1936 and 1937, the Congress made flood protection a national policy, and since then has undertaken widespread measures, such as reservoirs and dikes, to protect flood plains from overflow and to enhance the ability of land surfaces to absorb and so to retard the flow of storm rainfall. Nevertheless, the national damage toll continued to increase because building on flood-plain lands was not controlled. Current national policy is to publish maps that show the lands which are prone to flooding and to encourage local communities to enact programs that will control the misuse of flood plains.

LAND PHASE

Hydrologists, foresters, and other scientists have studied ways to control excessive evaporation or plant transpiration in order to increase the amounts of usable water without impairing the ecology of the region. If evaporative demands consume $14\frac{1}{2}$ inches of the 15 inches of annual precipitation that fall on a semiarid region, only $\frac{1}{2}$ inch is left. However, if evaporation could be reduced by a small amount, say by $\frac{1}{4}$ inch, the water supply would be doubled.

This could be done by using plant covers that require less water—for example, shallow-rooted plants instead of shrubs. Selective cutting of trees, especially those along streams or with direct access to shallow ground water, is another method. Still another way is to use chemical surfactants (cytel alcohols) that form a micro-molecular film and so retard transpiration from the stomata of plant leaves. These surfactants are sometimes applied by spraying and sometimes by being introduced into the soil, from which the plants absorb them with water.

CHANNEL PROBLEMS

Water, both liquid and solid, is an agent of weathering, of erosion, and of transportation. Sedimentary rocks are the products of erosion and deposition; rivers are authors of their own geometry in a well-adjusted balance, increasing in width, depth, and velocity, in that rank-order, as they move additional water from their headwaters to the sea. A river not only must drain its catchment but must convey the products of erosion as entrained sediment. Unlike continuously flowing water, the sediments may move in a process of deposition and scour in which, over long periods of time, the valley sediments are reworked as the river twists and turns—a natural movement known as meandering. This natural process means that the bottom lands and, especially, the flood plains are only lent, so to speak, by the river. However, whenever any change in the regimen of flow—such as is caused by diversions for water supply, urbanization, or dam construction—creates changes in the channel, those affected by that change are deeply concerned. Towns and cities, especially the growing suburban communities, are becoming anxious as the brooks that once gave charm to residential vistas are turning into scoured, storm-water channels—a consequence of missed opportunity to plan development in accord with principles of channel equilibrium (Figure 4).

The major duty of water is to transport waste substance and waste heat. Although man is the great polluter, streams in nature, even those that drain unpolluted areas, are not clean in every chemical or biological sense. Many streams, especially in the drier climates, are so loaded with salts that they are unpalatable or can be used only for industry or irrigation. Streams in areas of loose soils carry heavy loads of sediment and are nearly always muddy and roiled. Even remote mountain streams may contain harmful microorganisms in excess of the standards used for public supplies.

When a river receives an oxidizable waste from homes or factories—sewage, for instance—the oxygen in solution is consumed. The river water then has what is called

FIGURE 4 *This devastation resulted when a flash flood on Rapid Creek broke dams and deposited debris in what had been the bottom of Canyon Lake, Rapid City, South Dakota.*

33 *Water on the Land*

a saturation deficit, which must be restored by dissolving oxygen from the air. But there is a limit to this capacity beyond which water becomes putrescent. The problem can be handled in one of two ways—by treatment or by increasing the river's ability to assimilate oxygen. In the treatment method, wastes can be oxidized in a plant built for the purpose. The assimilative capacities of the river can be improved by building reservoirs to increase the flow during critically dry periods, and by such means as artificially aerating the streams, thus increasing their oxygen content.

In 1930, pollution was not a serious national problem. With a natural capacity to assimilate the wastes of 40 million persons, only 42 per cent treatment was necessary on a national scale. No treatment at all was needed if the oxygen assimilation capacity was doubled by augmenting low river-flow with water that had been stored in reservoirs during peak periods. By 1970, however, the pollution load created by a growing population and an expanding industry had tripled, so 85 per cent treatment was necessary, or 65 per cent if the assimilative capacity was doubled by reservoir storage. It is obvious that treatment has become a necessity, and flow-augmentation, which once was highly effective, can do the job no longer.

A new threat is thermal pollution. This is inherent in processes used to convert heat (whether nuclear, coal, or oil) to the mechanical energy needed to spin electric generators. It is rooted in the Carnot principle, which states that all heat cannot be transduced to mechanical energy. For instance, the spent hot steam of a steam engine or a turbine no longer is available for "work" and must be discharged to the air, which it heats, or condensed to water, which is also hot. Of the 0.75 pound of coal (or its equivalent) burned for each kilowatt hour of electricity produced, about half of the 10,000 British thermal units (BTUs) it contains is converted to electricity and the other half is removed by the coolant-water. Nonthermal methods or ways to convert chemical or nuclear energy directly into electricity may resolve or reduce the difficulties, some aspects of which are discussed in other chapters in this book.

WATER NEEDS VS. WATER REQUIREMENTS

When a city encounters shortages at the pumps, when the reservoir begins to run dry, the city fathers plan to enlarge the supply. Projections are made of the needs for water during the next 20 years by estimating the growth in population and industry and multiplying this estimate by a per capita use, increased to allow for ever-broadening uses of water, from air conditioning to swimming pools. The new scheme therefore calls for increasing the supply by bringing in new wells, building additional and larger reservoirs, and utilizing all of the technologies of water supply (Figure 5). When the new supplies are brought in, the demand increases further, creating new requirements, and the cycle begins again.

Many of the uses and abuses of water stem from its being considered a "free good"—there for the taking. Historically, this position was tenable. Today, however, the privilege is too freely interpreted, and some restraint—say higher price—may be needed to moderate the demand.

WATER EVERYWHERE

This very sketchy outline of the role of water—its physics, its circulation in the hydrological cycle, its use and abuse, its beneficent and hostile phases, its technology and its economics—only illustrates the scope of water science and water engineering. Its diversity is reflected in the kinds of people concerned—hydraulic engineers, meteorologists, geologists, limnologists, economists, foresters, agronomists, city planners, politicians, and, in his own way, the water-user himself. It is not the sole province of any single science or profession.

Here, perhaps more than in any other area today, science and public policy are entwined and, often, em-

FIGURE 5 *Hoover Dam, second highest in the United States, dams the Colorado River, forms Lake Mead (background) in northwest Arizona and southeast Nevada. It was the first major multipurpose facility: it provides irrigation, flood control, electric power, navigation, water supplies, and river regulation.*

battled. It comes down to one thing: if man is to have stable supplies of water for his many needs, he must be prepared to understand the problems—and to pay for the solutions. Irrevocably, man's existence on this earth is tied to water, and the same substance ties man to all forms of life.

CHAPTER 3 LOUIS J. BATTAN

The Earth's Atmosphere

...The sky gives and the earth receives....all things are mingled and all things combine, all things are mixed and all unmixed.... Zosimus, ca. A.D. 300

THE FIRST TWO CHAPTERS of this book have discussed the beginnings of earth and the importance water has played in its creation and existence. A second essential for life as we know it is the air we breathe. It sustains life, but can also take it away. It is ours to use wisely or to pollute unthinkingly. It can be a force whose power man fears and whose structure and control he is only beginning to understand.

The country of Bangladesh is no stranger to misfortune. Throughout the centuries it has been visited regularly by natural disasters of the most tragic kind But nothing in the recent past had prepared its people for the cyclone of November 12, 1970. This violent tropical storm swept in from the Indian Ocean on the crest of a high tide. The hurricane winds pushed the sea ahead of it in a storm surge 20-feet high. It flooded the low-lying islands and coastline, leaving behind incalculable destruction. Part of the storm's toll: 200,000 confirmed deaths and 50,000 to 100,000 missing; 400,000 homes damaged; almost 100,000 fishing boats destroyed.

The Bangladesh cyclone was one of the deadliest of all time, but its weaker counterparts happen every year in many places. These storms form in the layer of air that we call the atmosphere. Much of what we know about it has been learned only in the last few decades.

Since the middle 1940s, meterologists have used radar and instrumented airplanes to observe the detailed structure of thunderstorms, tornadoes, and hurricanes. Chemists and physicists have unraveled many of the mysteries of the formation and structure of clouds, rain, and snow. Studies of the behavior of fluids has brought about more realistic models of air motions and storm formations in the atmosphere. High-speed, electronic computers have made it possible to solve the equations of atmospheric motion. Sophisticated remote sensing systems carried on earth-orbiting satellites have yielded vital information about cloud, wind, and temperature patterns over most of the globe.

Only in recent years have scientists begun to study the earth's atmosphere as a single fluid system. As the first Apollo headed for the moon and the astronauts looked back from thousands of miles deep in space, they viewed for the first time—and photographed for all to see—the Planet Earth as it really is (Figure 1). They saw a nearly spherical body composed of continents and oceans and coated with an almost invisible, thin layer of atmosphere. Within this layer, over various parts of the globe, were cloud streaks and the spiral-shaped clouds of cyclones.

The earth's atmosphere is composed mostly of nitrogen (about 78 per cent) and oxygen (about 21 per cent), plus a mixture of many other gases. The relative amounts of the abundant gases are about the same up to altitudes of some 80 kilometers (about 40 miles), but certain gases, such as water vapor, decrease in concentration as they go higher. Ozone, which exists in small but important quantities, is found mostly at altitudes between 20 and 50 kilometers (12 and 31 miles, respectively). It absorbs ultraviolet radiation from the sun and helps to protect us from severe sunburns and skin cancer.

The atmosphere is commonly divided into layers (see Figure 2). The lowest one, through which the temperature decreases with height, is called the troposphere and

FIGURE 1 *This photograph of the earth from 22,300 miles in space was sent to the ground by the National Aeronautics and Space Administration satellite ATS III. A tropical storm (bottom center) has a cold front extending into Argentina. The line of clouds starting from left center to the northeast is a major weather pattern over Central United States, extending from Mexico to the Great Lakes and moving eastward.*

is some 10 kilometers thick, or about six miles. The layer above it, through which the temperature is constant or increases with height, is called the stratosphere. It extends to about 50 kilometers (30-plus miles) near the top of the ozone layer. There are several other layers above the stratosphere through which the temperature alternately decreases and increases with height.

Oxygen and nitrogen, of course, make our kind of life system possible. But what keeps those gases in the atmosphere? Why don't they rise, escape into one or more of the layers above, and eventually leak through into

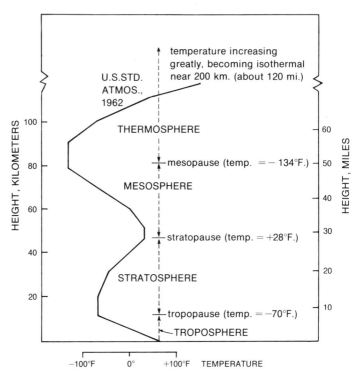

FIGURE 2 *Profile of the atmosphere. Air density decreases as height from the earth increases.*

outer space? A highly simplified explanation is that the molecules and atoms of the various elements in our atmosphere are constantly on the move—each type "flying" at a specific average velocity. Collisions are inevitable, and slow-moving particles check the speed of faster-moving ones when they hit each other. Atoms and molecules in the upper atmosphere—particularly the lighter ones—do escape at a very slow rate. However, oxygen and nitrogen are heavy, and only minimal amounts leave the atmosphere. These can be replenished by chemical activities of plants on earth. Hydrogen and helium, which are light, have escaped constantly, so relatively small amounts remain in the air we breathe.

Now, if the earth were as hot as the sun, the velocity of

all atmospheric molecules would be speeded up considerably. In that case, oxygen and nitrogen, like hydrogen and helium, would have risen above the atmosphere at significant rates during earth's long history and very probably we would not be telling this story.

WATER FOR A THIRSTY WORLD

In many respects, the troposphere is of most interest to us all because it contains the air we breathe and is the region in which clouds form and produce the rain and snow necessary to maintain life on earth. In Chapter 2, Dr. Langbein discussed various aspects of the water problem, one of which is getting increasing attention from atmospheric scientists: What causes droughts and can anything be done about them? Droughts occur when atmospheric conditions inhibit the formation of clouds. If there are no clouds, there will be no rain or snow. But why are there no clouds? Meterologists have a ready answer. They can show that droughts occur because abnormalities in the wind patterns cause the air to sink over regions where it would normally be expected to rise. When air sinks, clouds generally cannot form. But you might ask, "Why does the air sink?"

This question brings the discussion to an examination of a fundamental characteristic of the atmosphere of any planet—the so-called *general circulation*, which refers to average motions of the air over the globe. The characteristics of the air motions are important because they determine the weather and the climate in any given part of the world.

GENERAL CIRCULATION OF THE ATMOSPHERE

The general circulation of the atmosphere depends on a number of factors. Some are obvious, others are not. First, the energy to drive the motion of the air comes from the sun. Its radiation warms the equatorial regions more than the polar regions, and the resulting temperature differences play a crucial role in determining the properties of the over-all wind patterns. A second important factor is the speed at which the earth rotates. Third is the configuration of continents and oceans. Ocean water temperatures change slowly and the currents transport large quantities of heat. On the continents, mountain ranges interact with the winds. All of these items, plus a few others, are interrelated in a complicated way. For example, as warm, dry air blows over the oceans it can cause waves, and the air's heat is transferred to the ocean as water vapor from the waves evaporates and moistens the air. In other circumstances, the sea supplies heat to the cooler air above it.

It is now possible, to a certain extent, to take into account the various factors that influence the general circulation and to construct a mathematical model of the earth's atmosphere. Such a model, when programed into an electronic computer, will yield pictures of the earth's atmosphere that are quite similar to those on weather maps of the earth. Such maps are based on observations from hundreds of weather stations, but at the present time they leave much to be desired. In many parts of the globe, especially over the oceans in the Southern Hemisphere, observations are sparse or nonexistent. Fortunately, modern techniques of indirect probing of the atmosphere by means of earth-orbiting weather satellites (see the chapters by John Pierce and Arthur Schawlow) are beginning to fill the gaps in the traditional weather-observing network. An international research operation known as the Global Atmospheric Research Program is currently engaged in designing a system to observe the entire atmosphere adequately, to develop a much-improved understanding of the way in which it functions, and to construct a mathematical model of the atmosphere of the entire earth.

Weather maps of the Northern Hemisphere commonly show a continuous flow of air around the globe in the form of meandering currents (Figure 3). Meterologists refer to the strongest currents as "jet streams." The wavy patterns are characterized by regions of low pressure called troughs and regions of high pressure called

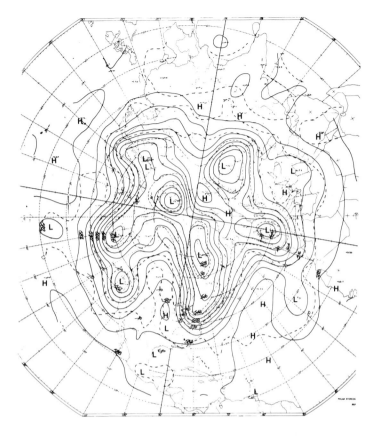

FIGURE 3 *This is a 500-millibar map of the northern hemisphere that shows height contours, at an altitude of about four miles, in tens of meters (10 meters = 32.8 feet). A millibar is a measure of atmospheric pressure. Temperatures are shown in degrees Celsius (0° Celsius = 32° Fahrenheit). Winds at the level at which the pressure is 500 millibars blow in directions nearly parallel to the contours. The closer the lines, the higher the wind speed. The map was made on June 10, 1972.*

ridges. Commonly, the low-pressure troughs are regions of rising air, cloudiness, and rain or snow, while high-pressure ridges are regions of sinking air and fair weather. These relationships between wind patterns and weather are the keys to accurate weather forecasting. The chief practical aim of the Global Atmospheric

Research Program is to develop techniques that can predict the wind patterns and the weather for one to two weeks in advance.

If you examine weather maps over a 10- or 20-year period, you find preferred positions for the troughs and ridges. For example, in the winter, the eastern United States commonly is under the influence of a persistent trough and, as a result, the weather generally is wet and cold. As you would expect, regions in which the air blows from north to south usually are colder than those dominated by air that blows from south to north.

THE CLIMATE AND ITS CHANGES

Speaking loosely, the climate of any place can be said to be the weather averaged over a long period of time—some tens of years. Although changes of climate are slow, they go on for long periods and therefore large changes can occur. Over geological time, ice ages were associated with cool, wet periods separated by intervals of warm weather. The last true cold spell experienced on earth, and often called the "little ice age," lasted for some 200 years and ended about 1840. This was followed by a warming trend, which continued for about a century. Since 1940, earth has been cooling off again—irregularly at the beginning, but more sharply since 1960.

Scientists have been particularly concerned about the possible effects of human activity on the climate changes during the last century. We know, for instance, that carbon-dioxide concentration increased from about 295 to about 322 parts per million of air from 1870 to 1970. Largely responsible for this pronounced increase are our automobiles, power plants, smelters, factories, and even homes, because carbon dioxide is formed when fossil fuels are burned. Carbon dioxide is not toxic, but it may be important in determining the earth's climate because it interferes with radiation transfer in the atmosphere. These same sources contribute other pollutants, such as sulfur dioxide, nitrogen dioxide, and carbon monoxide. The amounts differ widely from place to place, and sometimes exist in high enough concentrations to be toxic.

Modern civilization is also responsible for huge quantities of solid and liquid particles liberated into the air. Most particles do not remain in the lower atmosphere for long periods—perhaps a week or two, on the average. Large particles fall to the ground; smaller ones are removed by precipitation. On the other hand, tiny particles in the stratosphere may stay there for a long time—more than a year. Particles also may serve as nuclei for the formation of clouds, which, in turn, can reduce significantly the amount of solar energy absorbed by the atmosphere.

There is no doubt that the air quality and climate of large cities generally are different from those in the nearby countryside. For example, cities have higher temperatures. Some evidence has been gathered indicating that the rainfalls downwind of Chicago and St. Louis are greater than they would be if those cities did not exist. In 1971, an extensive network of weather stations was established around St. Louis in order to measure the extent to which the urban area was affecting the weather and climate. The results are not yet in, but they surely will pose questions whose answers could affect all of us in the years to come.

When land is cleared for highways and buildings or the ocean waters are covered with a layer of oil spilled from ships or leaking from oil wells, the rates of exchange of energy, moisture, and momentum between the air and the earth change. These factors, and others as well, influence the circulation of the atmosphere in complicated, but relatively minor, ways.

The effects on the atmosphere of all these various forms of environmental pollution have not yet been evaluated in a satisfactory way. For instance, there are still conflicting views over whether the concentration of particles in the atmosphere has increased as a result of man's activities. After all, nature itself contributes in major ways to such pollution—volcanic eruptions are a major source of high-level dust. However, if man is indeed causing weather and climate to change, it is crucial to life on earth that we find out as quickly as possible. This can be done only by collecting appropriate measurements of the properties of the environment and by developing realistic mathematical models of global climate that can take into account the complex interactions among the major relevant factors.

CHANGING THE WEATHER

During some years, for reasons that are not clear, the entire pattern of troughs and ridges as seen on weather maps shifts longitudinally from its normal position. For example, the trough normally found over the eastern United States may become established over the western part of the country. When that happens, the West experiences an abnormally wet winter and perhaps flooding, and the East has unusually dry conditions. In extreme cases, especially if the pattern repeats itself several years in a row, the eastern states experience a drought.

In recent years, many people suffering the effects of droughts (Figure 4) have sought ways to overcome them by means of one of the newest, and perhaps most controversial, of the emerging technologies—cloud seeding. It must be recognized, however, that for it to have any chance of success, special types of clouds must exist.

Most clouds form by the condensation of water vapor, which rises and cools. When updrafts carry the cloud to higher altitudes, the tiny water droplets that constitute the cloud become progressively colder. Surprisingly, even when the temperatures fall below freezing, the droplets often do not freeze. Such *supercooled* clouds may remain in the liquid phase until they ultimately evaporate without ever producing rain or snow. On the other hand, if some ice crystals are introduced into a supercooled cloud, it changes rapidly as the ice crystals grow and the water droplets evaporate. Sometimes, the crystals become so large that they fall out of the cloud in the form of snow. In the process just described, the ice crys-

FIGURE 4 *During severe droughts, vegetation dies, leading, in many places, to blowing sand that completes the devastation.*

tals could be said to have "seeded the supercooled cloud." In nature, ice crystals form on minute particles of a special type known as *ice nuclei*. Sometimes, though, there is a deficiency of natural nuclei, and a supercooled cloud may never be converted to ice crystals by natural processes.

This theory of snow formation was developed in the early 1930s, but not until about 1946 was it realized that the phenomenon offered a means for man to change clouds and precipitation. Vincent J. Schaefer, who was investigating the nature of supercooled water drops, discovered that ice crystals formed when Dry-Ice fragments fell into a supercooled cloud. Shortly thereafter, one of his colleagues determined that the crystal structures of silver iodide and lead iodide were similar to those of ice and that they were effective ice nuclei. Over the years, Dry Ice and tiny, silver–iodide particles have been the most widely used substances for artificial cloud seeding.

A scientific concensus has been growing about the ability of cloud seeding to increase or decrease rain or snow by significant amounts. The evidence is strong that, in certain meterological circumstances, precipitation can be increased substantially—some tens of percent. In other circumstances, the seeding of ice nuclei might cause precipitation to decrease by about the same amount. In other circumstances, seeding has no effects. The most convincing data to support these assertions have come from carefully conducted tests in which clouds were seeded on a random basis and precipitation was measured.

It is important to remember that unless there are clouds of a particular type, cloud seeding is useless. For this reason, cloud seeding cannot "break a drought." As mentioned earlier, droughts are caused by characteristics of the air flow that suppress the formation of clouds. Of course, on the few occasions during a drought when clouds do occur, cloud seeding might cause them to yield more rain or snow. On the other hand, one must be concerned that the seeding might actually decrease the precipitation and worsen the drought, although cloud seeding cannot be blamed for causing a drought. Notwithstanding the increasing optimism about the potential of cloud seeding, much more must be learned and serious legal questions resolved before it can be regarded as a practical technique in the catalogue of the water engineer.

CONTROLLING HURRICANES

Some meterologists have become hopeful that cloud seeding might do much more than modify clouds and precipitation. There are reasons to believe that it might mitigate the hazards of violent storms, such as tropical cyclones and hurricanes (Figure 5).

During the last decade or so, the United States Government has undertaken Project Stormfury to investigate the nature of hurricanes and try to learn how to weaken them. A number of experiments have been conducted in which hurricane clouds were seeded heavily with ice nuclei. The most satisfactory tests were carried out in

FIGURE 5 *The swirling clouds at top right are Hurricane Inez, centered off the west tip of Florida on October 5, 1966. The photograph was taken by ESSA 3 Weather Satellite on October 5, 1966.*

August 1969, when Hurricane Debbie was seeded on two separate days. At the start of seeding on August 18, the maximum wind speed at the flight altitude of 12,000 feet was about 113 miles per hour, but it fell to 78 miles per hour after the seeding. Two days later, when the storm had reintensified and wind speeds reached 114 miles per hour, the storm was seeded again. After that second seeding, maximum winds were about 15 per cent less than at the beginning of the test. In addition to this experimental data, theoretical studies by means of mathematical models also indicate that, under some conditions, maximum hurricane winds might be reduced significantly by means of cloud seeding.

The evidence collected so far, although limited, is encouraging, but more experiments are needed to double-check the existing methods. For instance, we must find out if ice-nuclei seeding, under certain special circumstances, could cause the unexpected and produce an increase in the intensity of the storm or possibly change its path. Needless to say, enormous potential benefits to society would be afforded by a technique that reduces the violence of hurricanes. We must learn as quickly as possible if, and how, it can be done. And here, too, legal questions are involved. Let us hope we do so before another devastating tropical cyclone moves into Bangladesh or any other part of the globe.

CHAPTER 4 RICHARD H. JAHNS

The Solid Earth

The stability of the solid earth is instability itself.
Alexander Winchell, in *Walks and Talks in the Geological Field,* 1886

OUR EARTH has long been recognized as a rather lively body, quite apart from its gross motions as a member of the solar system. Its surface, for example, is constantly being reshaped by water, wind, and ice acting under the influence of temperature differences and gravity, at times in spectacular and even catastrophic ways. The situation at depth also is far from static, as we are reminded from time to time by earthquakes and volcanic eruptions. Indeed, the entire history of our planet seems to have been characterized by massive transfers of energy and materials.

The causes, nature, and effects of these transfers have been under study for a long time, but the centuries of man's observations and measurements amount to no more than a tiny fraction of earth's total history of several billion years. Nonetheless, this complex history is at least partly understood today, an achievement deriving from two major sources of information—processes and products of the present, and rocks that are products of the past. Thus it has proved feasible to feel the pulse of current earth activities, and then to apply the findings to a reading of the available geologic record. The scale on which earth can be examined varies from the atomic to the cosmic, and whether the approach is by simple observation and inference or by a lengthy series of carefully reasoned investigative steps, it can provide excitement akin to that of the finest detective story.

Only a century ago, when our government undertook geographical and geological surveys of the western territories, the scientists on these expeditions were faced with serious problems. Vast expanses of country were to be covered in a limited time, transportation was difficult, and physical survival was by no means assured. In a geological context, little was known of the region, and certainly much less was known about the earth as a whole than is now the case. Yet these men did have some advantages, for at that time much already had been accomplished in geologic studies elsewhere. Clear distinctions, for example, by then had been made between igneous rocks, formed through crystallization of molten silicate materials, and sedimentary rocks, formed mainly from solid materials deposited or precipitated in bodies of water. And it had been recognized that rocks of either kind, if deeply buried under conditions of high temperature and pressure, could be recrystallized and even compositionally changed in the solid state to form metamorphic rocks. As a further bonus, some earlier misconceptions about these different rock types and their interrelationships already had been corrected.

The scientists were remarkably efficient in their reconnaissance studies. Fortunately, they were able to begin, in many areas, with distant "big looks" at vistas of well-exposed rocks, which enabled them to observe and to record broad-scale relationships and better to decide where specific visits should be made for rock identification, sample collection, and closer study. In effect, they were able to see and appraise the forest before getting directly involved with the trees, whether the main feature was the front of the Uintah Mountains or the canyon of the Yellowstone River. Once they had the general situation in mind, they moved in for detailed study at selected spots. Some of this work involved the collection of fossils representing ancient life and hence valuable for rela-

FIGURE 1 *The Grand Canyon. In this great opening, cut during the past few million years by the Colorado River, one can observe lengthy chapters of earlier geologic history. Dominating the scene are nearly horizontal layers of sedimentary rocks representing ancient deposits of several kinds, and below them are older sedimentary strata that have been tilted and otherwise deformed. Exposed at the bottom of the canyon and along the inner gorge of the present river are igneous and metamorphic rocks more than a billion years old.*

tive age-dating of the host rocks, and much of it was focused on descriptions of the rocks themselves.

Nowhere is the value of an initial "big look" more evident than at Grand Canyon, where the Colorado River has cut its way downward through thousands of feet of sedimentary rocks that reflect more than 300 million years of earth history (Figure 1). Handsomely exposed on the irregular canyon walls are nearly horizontal layers, or strata, of contrasting thickness, color, texture, and composition. They can be traced by eye for miles, their continuity broken in only a few places by offsets along near-vertical faults. This great section of rocks, representing a variety of ancient sediments deposited mainly in shallow seas, is underlain deep in the canyon by older sedimentary layers that are inclined at moderate angles. This older section, once essentially horizontal, was deformed by tilting and then was partly removed by erosion prior to accumulation of the younger strata. Beneath its remnants, the narrow inner gorge of the canyon exposes igneous and metamorphic rocks that reveal still earlier chapters of earth history.

No doubt the geologic explorers of a century ago could have profited from even broader views of western terrains, but this was denied them because they were earthbound. Not until man was able to invade the atmosphere in balloons and airplanes could he gain the advantage of seeing more and more from one place and of recording such views photographically. In recent years, this capability was extended further through programs of space exploration, until astronauts were able to observe and photograph the entire earth at one time. On this score, it is interesting to note that geologic exploration of the moon necessarily has followed an opposite course through the centuries. The earliest observations, earth-based and hence very broad, were made first with the unaided eye and then through telescopes with progressively increasing detail. Later came photographs and other imagery from space vehicles traveling nearer and nearer to the lunar surface, and even a chemical analysis of lunar material recorded from an unmanned spacecraft. Finally, the Apollo program provided six geologic field trips, by far the longest ever made by man. When astronauts Aldrin and Armstrong first set foot on the moon in 1969, thereby opening it up as a new frontier, they had much better knowledge and far more sophisticated equipment than did those earlier explorers of the American West. Many of their basic working objectives, however, were little different; they, too, were concerned with rock identification, sample collection, and closer study.

Here it is pertinent to mention a few aspects of lunar

geology before returning to an examination of the solid earth. First, the moon's surface, like that here on earth, is far from uniform or monotonous. It includes mountain ranges, plateaus, valleys, and some deep canyons, and it is particularly featured by broad basins and by craters of many sizes. Some of the depressions may prove to be of volcanic origin, but most of them evidently were formed by solid materials striking the lunar surface. Indeed, the moon appears to have undergone bombardment on a terrible scale early in its history, probably more than three billion years ago, and to have absorbed countless lesser blows since then. So much energy was involved in the largest of the impacts that craters tens of miles across were formed, parts of the lunar crust were shoved upward to form ridges and even mountains, and other parts probably were melted to form lavas that filled some of the major depressions.

Second, the lunar rocks thus far examined are similar to rocks well known on earth, but all are remarkably fresh and unaltered. Most contain abundant glass that reflects melting from the energy of meteorite impacts, and their surfaces typically are pitted by tiny craters. The most widely encountered rock types are breccias, which consist of angular rock fragments in a matrix of finer-grained debris that includes bits of rocks, minerals, and glass. Some of the breccias may be of volcanic origin, but most seem to represent fallout from various impact sites. No deposits formed in bodies of water have been found thus far, and most investigators currently agree that the lunar surface has had an essentially waterless history.

Third, the exposed lunar rocks are very old, with determined ages ranging from three billion to slightly more than four billion years. But younger rocks may be present, as six field trips hardly suffice for sampling what the moon has to offer. Seemingly this body is not wholly dead in terms of internal processes, as attested by increases in temperature with depth, some evidences of relatively young volcanic activity, and a modest number of "moonquakes" recorded by seismic instruments emplaced during the Apollo program. Yet the present sum of investigations suggests that, for a very long time, changes on the lunar surface have been imposed mainly from external sources. Continuing bombardment by meteorites of various sizes has stirred and rearranged the surface materials over a wide range of scales, an extremely slow process referred to as "gardening," and the larger impacts have formed correspondingly larger craters. Major features of the lunar scene have been progressively softened, partly filled, interrupted, or otherwise modified with the passage of time, but they nonetheless remain extraordinarily well preserved for their ages.

Why should such survival be regarded as extraordinary? Because it is just that, when viewed in the context of the earth. Here on our planet we can see no craters or mountain ranges three billion years old or even one billion years old, and relatively few impact craters of any age. Nor have we yet discovered rocks older than about 3.5 billion years, or more than a very few that are older than two billion years. We have found little terrestrial glass older than about 50 million years, and little that is of impact rather than volcanic origin.

Do these and other differences mean that earth and moon have been exposed to fundamentally different external conditions as bodies in the solar system, and that earth has been largely spared, for example, the bombardments moon has received? Not necessarily. A much more likely explanation for the observed differences lies with some basic contrasts between the two bodies themselves. Earth has long had an atmosphere, abundant water, and organic life, whereas moon has not. And processes operating beneath earth's surface have made it much more active than the moon through most of its history. Were we to take a space trip and return to earth like the Apollo astronauts, frictional heat would remind us of its atmosphere when we reentered it, we would splash down into some of its water, we would be greeted by people, and we would have earlier seen expressions of its liveliness from many points on our long journey.

Exposed solid parts of Planet Earth consist mainly of silicate and aluminosilicate materials in the form of sediments and rocks. Only eight elements—oxygen, silicon, aluminum, iron, calcium, sodium, potassium, and magnesium—account for about 99 per cent of these exposed materials, with oxygen contributing nearly half of their total mass. This element is even more dominant in terms of volume, so that we not only breathe it from the atmosphere but also reside upon the much larger amounts of it that are tied up chemically in the oceans and ground.

Many rock-forming substances are unstable at ordinary surface temperatures, especially in the presence of water, carbon dioxide, organic acids, and free oxygen. Given enough time, therefore, even the strongest and hardest rocks can be dissolved or chemically altered to form soils and other materials. Such decay has been essentially absent from the lunar surface, which has lacked the necessary water and atmospheric agents. Similarly, even the most ancient glasses on the moon have remained fresh, whereas terrestrial glasses have been slowly devitrified (crystallized) or altered in the presence of water and other fluids.

The earth's surface, unlike that of the moon, also has been continuingly modified by mechanical processes quite unrelated to the impacts of meteorites. Rocks have been broken up, for example, by the actions of plants, man and other animals, freezing water, and wholly internal stresses, and the fragments have been carried away by water, ice, and wind to other localities. Thus a turbid Colorado River, charged with enormous quantities of sand and silt derived from a Painted Desert or a Rocky Mountain front, is able to carve out a Grand Canyon and deposit its load to form a delta in a Lake Mead or a Gulf of California. Or the torrential waters of a rainstorm in a Sierra Nevada or Wasatch Range mobilize previous accumulations of talus and other rock detritus to form bouldery slurries that pour from canyon mouths and devastate adjacent valleys (Figure 2). Or loose materials are gathered up by the fierce winds of a Sahara or Colorado Desert, and then are dropped to form extensive belts of dunes. Or, looking back a bit in time, a great moving ice cap of 20 thousand years ago, having thoroughly scoured the ground surface over millions of square miles in a Europe or North America, slowly dissipates to leave irregular blankets of debris across the landscape near its former margins. Small wonder that earth's surface, as a scene of relatively rapid action, bears so few scars of meteorite impact. The craters sooner or later disappear from the scene as materials of their rims and walls decay, disintegrate, and are either carried away or buried beneath younger deposits. And the landscape as a whole simply doesn't "stay put" long enough for extensive "gardening" by those meteorites large enough to survive the frictional heating of travel through our planet's atmospheric envelope.

If the terrestrial surface is being so vigorously modified in ways that we can observe and measure, we might well ask how long such activities have been going on. The answer is given by the many kinds of exposed rocks that range in age from yesterday to points far back in earth history, and that represent earlier episodes of erosion, sediment transport and deposition, glaciation, volcanic action, and so on. Too, some products of ancient meteorite impact can be recognized, even though most of the related craters have long since been erased or buried. But all this must prompt another question, for most surface processes tend to transfer materials to lower and lower positions, either on land or within water-filled basins. If these processes tend to be very rapid as measured against geologic time, why shouldn't earth's surface have been reduced long ago to a uniformly low level? And if the available evidence does suggest that some such reduction has occurred over broad regions at various times, why do so many high plateaus and mountain ranges nonetheless exist today?

The answer is to be found beneath us, where matter and energy have long been shifted about in profound ways. As manifestations of these activities, the outermost parts of the earth have been repeatedly moved in both vertical and lateral senses, squeezed and ruptured in many styles, and invaded from below by molten materials. The various kinds of disturbances have ranged widely in their scale, location, and timing, and so complex has been the interplay among factors of construction, modification, transfer, and destruction that remarkable contrasts can be recognized in the geology of adjacent areas and regions. Thus, much of the Mississippi River basin is underlain by nearly horizontal sedimentary strata that have been no more than mildly disturbed and probably have remained near the earth's surface since they were laid down in shallow seas more than 250 million years ago, whereas strata of like age in nearby New England have been deeply interred, severely deformed, metamorphosed, and interrupted by numerous bodies of igneous rocks. To the north, over large parts of eastern Canada, such strata long since have been stripped away by erosion to reveal a much older and more complicated sequence of igneous and metamorphic rocks like some of those exposed along the inner gorge of Grand Canyon. In more recent geologic times, rocks less than 100 million years old have been pervasively crumpled, sheared, and metamorphosed in coastal California, where thick accumulations of still younger strata have been variously ruptured, offset, and juxtaposed against quite unlike rocks along large faults. And, as a last contrasting example among many that could be cited, a considerable variety of rocks has been concealed beneath vast outpourings of geologically young basaltic lavas in much of eastern Oregon and Washington.

Returning now to what lies farther beneath us, we must acknowledge certain limitations in examining the vertical dimension. Surface observations are essentially restricted to exposed parts of the continents and to numerous oceanic islands, which together account for slightly more than one-quarter of the earth's surface and provide an extreme vertical range of slightly less than six miles. A similar extreme reach for controlled sampling beneath the surface represents man's deepest drilling to date, but direct observations in the subsurface cannot be extended below approximately two miles, at the bottoms of the deepest mines thus far developed. Some samples have been carried to the surface from considerably greater depths by obliging volcanoes, but such materials cannot be labeled as to exact original location. Fortunately, much can be inferred about earth's interior by indirect means. Determinations of shape, radii, mass, gravity, and moment of inertia for this planet indicate that its average density is substantially greater than that of its outermost parts. They further impose certain restrictions on how density might vary with depth, thereby furnishing clues about compositional variations. With additional information derived from records of earthquake-generated waves, a reasonably satisfactory model of a layered earth can be developed (Figure 2).

The outermost layer is the crust, a complex solid but mobile skin three to nearly 45 miles thick. Its average composition probably is little different from that already noted for its exposed surface. The underlying layer, termed the mantle, is about 1,800 miles thick and accounts for most of earth's volume. It also is solid, but seemingly is capable of flowage under slowly applied stresses. Its outer parts, consisting mainly or iron- and magnesium-silicate minerals, are significantly denser than the crust. Its inner parts, which become progressively denser with depth, probably are different in some aspects of mineral composition. They may contain more iron, although the chemical composition of the entire mantle could be fairly uniform. Next beneath is the outer core, which is nearly 1400 miles thick and may well consist of iron and nickel, with lesser amounts of lighter elements. It evidently is liquid-like in its behavior, as it is incapable of transmitting earthquake shear waves. The inner core, with a radius of about 750 miles, is solid, very dense, and probably metallic in composition.

Two fundamentally different kinds of crust can be dis-

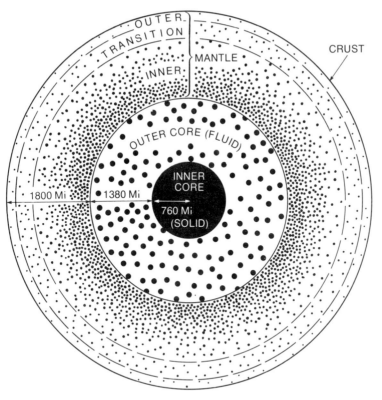

FIGURE 2 *Model of an internally-layered earth that is consistent with inferred variations in density and with observed behavior of earthquake-generated (seismic) waves. Most of the earth's crust is thinner than the outermost line at the scale of this section.*

tinguished on earth. One kind, which forms the continental masses distributed as platforms or plates in both hemispheres, is a heterogeneous mosaic of many rock types. It varies considerably in thickness from place to place, and extends to its maximum depths beneath the earth's great mountain belts. These belts mark the general positions of former subsiding basins in which enormous quantities of sediments were deposited. The resulting layered rocks, where most deeply buried and downbuckled, were metamorphosed and partly melted to form the root zones of growing mountain ranges. As the mountains were uplifted in an environment of crustal compression, large volumes of molten materials moved into their cores and crystallized there as granitic rocks. The most recently formed mountain chains, in general less than 70 million years old, can be correlated with much of the earth's present highest ground. Older chains have been reduced by erosion, but their complex roots are well exposed in several parts of the world.

The other kind of crust, which underlies the ocean basins, consists mainly of basaltic rocks and hence is richer in iron and magnesium, somewhat poorer in silicon. It also is much thinner and extends to more uniform depths. Like the continental crust, it has been very active through geologic time; its behavior, however, has been quite different. Studies of the ocean floors during recent years have revealed some truly remarkable features, among which are highly elongate ridges that are sites of new crustal additions. One of these ridges, for example, follows a sinuous course beneath the Atlantic Ocean, and a much longer one—apparently an extension of it—has been traced across the basins of the Indian and Pacific oceans from Africa to North America.

With discovery after discovery during little more than a decade of investigating sea-floor topography, seismic activity, heat flow, sediments, and rocks, a new and unifying concept of crustal behavior has emerged for testing by further studies. Called global plate tectonics, it features progressive and systematic separation, or spreading, of oceanic crustal plates along the axes of the sea-floor ridges, accompanied in these elongate zones by contributions of new basaltic material derived from the underlying mantle (Figure 3). As the new material solidifies and cools, it is weakly magnetized in accordance with the earth's existing field. But the polarity of the earth's field has reversed at intervals in the geologic past, so that successive past contributions of basalt from the ridges now can be mapped on the ocean floors as alternating broad stripes of rocks with contrasting magnetic polarity. By such means it has become possible to trace the gradual expansion of ocean basins, which apparently has occurred at brisk average rates of inches per year.

Widespread evidence in both the continental and oceanic crusts suggests that a single supercontinent existed as recently as 200 million years ago, and that it was progressively broken up into smaller continental masses that gradually drifted apart. In that way, the north Atlantic Ocean was born as North America was separated from Europe along a new spreading zone that now is identified with the mid-Atlantic ridge. Similarly, South America was separated from Africa with development of the ridge farther south. As an example of a much younger beginning, probably less than 5 million years ago, the Gulf of California was opened up where a spreading zone of the eastern Pacific Ocean basin cut into upon the North American continent. Today a narrow strip of relatively fresh oceanic crust with accompanying small volcanoes is exposed window-like between Sonora and Baja California. Any progressive separation of land masses that were once joined can provoke interesting thoughts about such diverse matters as intercontinental correlation of old mountain belts or mineral deposits, and the evolution of plants and animals that once lived together but became isolated on their respective moving rafts.

If sea-floor spreading provides a reasonable mechanism for the drifting apart of continental plates, and if the major oceanic basins have been enlarging during the past 200 million years, how has room been made for the added crustal materials? The most likely explanation for this question of space can be found along some of the continental borders, where oceanic crust evidently has been thrust beneath thicker continental crust. Good examples include the western margins of South America and Central America, the southern margin of Alaska, and the island arcs off the coast of eastern Asia. Not surprisingly, these belts of inferred underthrusting have corresponded rather closely with the positions of oceanic trenches and belts of strong earthquake and volcanic ac-

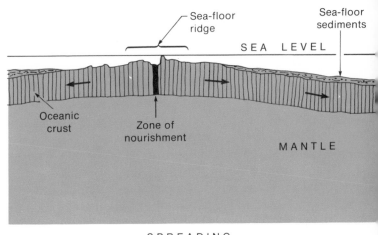

SEA LEVEL

Sea-floor ridge

Sea-floor sediments

Oceanic crust

Zone of nourishment

MANTLE

SPREADING

FIGURE 3 *Simplified sections illustrating the spreading of oceanic crust and the movement of oceanic crust beneath adjacent continental crust. Heavy arrows indicate general directions of slow movement. In section at left, new basaltic crust is supplied from the underlying mantle along the axis of a sea-floor ridge as sediments gradually accumulate on older parts*

tivities. The Alaska earthquake of 1964 and major volcanic eruptions in Central America have provided some of the testimony in historic times.

The driving force for gradual movement of sea-floor plates away from the zones of nourishment is widely thought to be furnished by cells of slow convective circulation in the underlying mantle. This and other elements of the global tectonics concept remain to be firmly established, just as some apparent inconsistencies still require resolution. At present, however, the concept presents an exciting opportunity to "put it all together" for a fraction of earth's history, and a challenge to look farther back in time for evidences of earlier interplay among crustal masses of continental and oceanic affinities.

This discussion of the solid earth can be concluded with a note on man's brief participation in its history. If this history were likened to a play that has continued for

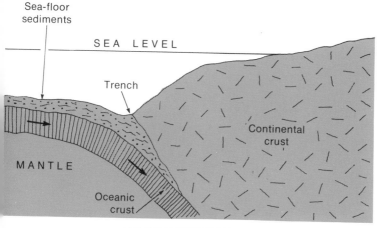

Sea-floor
sediments

SEA LEVEL

Trench

Continental
crust

MANTLE

Oceanic
crust

UNDERTHRUSTING

of the sea floor along flanks of the ridge. In section at right, oceanic basalt and its cover of sediments are shoved beneath the margin of a continental mass along a sea-floor trench. The boundary between oceanic and continental crusts, inclined at an angle of about 45 degrees, is a zone of major earthquake generation.

two hours, man would have appeared on the stage only about three seconds ago, yet his impact already has been felt. He has modified parts of earth's surface in many ways, with good results like the control of seasonal flooding and soil erosion, and with bad ones like the promotion of soil erosion and unwanted landsliding (Figures 4A and 4B). He has drawn heavily upon earth's nonreplenishable resources, most of which were formed long before he joined the action, and in many instances he may have expended them not wisely but too well. He has been slow to recognize the hazards or to avoid the risks of living on flood plains of rivers, edges of unstable cliffs, flanks of volcanoes, or traces of active faults (Figure 5), yet he has been sufficiently agile to survive such significant events as widespread falls of volcanic ash and major changes in sea level attending the growth and waning of continental glaciers. Indeed, it could be argued that he has proliferated and prospered as he has learned to obtain higher

65 *The Solid Earth*

FIGURE 4A *A seacliff landslide at Pacific Palisades, California. Sudden failure of weak, water-saturated sedimentary rocks sent a large tongue of debris across the coastal highway, which was being cleared when this photograph was taken in 1958. Remnants of older landslide masses can be seen along adjacent parts of the steep, seaward-facing slope, which has been slowly encroaching upon the settled area in background.*

yields from the soil, to discover and dig deeper in deposits of useful minerals, to harness new sources of energy, and to cope more effectively with his physical environment.

Man has inherited the earth, and he finds it in fundamentally good condition. It has plenty of thermal energy, some retained from its beginning times and some generated from radioactive constituents in its crust, so that he needn't worry even though he can't yet decide whether earth has been slowly heating up or cooling

FIGURE 4B *A moving landscape in the Palos Verdes Hills of southern California. The hummocky ground shown here reflects landsliding that has occurred repeatedly during the past million years, generally in directions toward the ocean. This ground also has been gashed by several stream-cut canyons. In 1956, when the photograph was taken, man already had developed parts of this attractive area for residential purposes, and a new access highway was under construction (upper right). Materials from the large roadcuts were deposited as fill on adjacent parts of the ancient landslide complex, and the added mass contributed to sudden renewal of slippage that involved much of the ground between the new highway and the ocean. Many of the homes shown in this view were wracked and destroyed by the irregular slide movements, which have continued to the present time.*

FIGURE 5 *A breached reservoir in Los Angeles, California, that reflects unfortunate interplay between man and the local geologic scene. This facility was constructed across the surface traces of several small faults that appeared to be inactive. Movements of several inches occurred along at least one of these faults, however, and ultimately led to failure of the reservoir wall and catastrophic flooding of a nearby residential area in 1964.*

down. Nevertheless, it is a finite body with finite material resources, and he has more homework to do if he is to last as long and thrive as well as some other species of record. It is to be hoped that he can continue to learn more and more about the solid earth, and that he can apply the additional knowledge toward increasing wisdom in his future behavior.

LIFE

As life arose on our planet and man evolved from lower forms, he was set apart from all other living things by his brain. The first chapter in this section describes some aspects of that marvellous machine, and points out that without it man would not be aware of problems—or of curiosity, the burning need to probe the unknown. Some of our best brains today are working on two questions of incalculable importance to the entire world. What is to be done about the exploding population? How are the millions to be fed? These are basic questions of life itself—its beginning and, perhaps, its end, and they are closely interrelated. As the population grows, it must be fed with food that will insure healthy bodies and healthy brains that can discover more ways to handle the population, which must be fed...etc. At least, that is a pessimistic viewpoint. But we need not live on such a merry-go-round. Research already has found many answers. Perhaps the next step is up to us.

CHAPTER 5 SIR JOHN C. ECCLES

Neurosciences: Our Brains and our Future

We are a part of nature, and our mind is the only instrument we have, or can conceive of, for learning about nature or about ourselves.
C.H. Waddington, in *The Nature of Life*

MORE THAN EVER BEFORE, brain research is being recognized as the ultimate scientific challenge confronting mankind. Ever since his realization of his existence, man has been trying to understand what he is, the meaning of his life, and how to conquer and control the land, air, and water that are the bases of his existence on this small planet. No one can argue against the brain being central to the life of man. But for man's brain, no cosmological or environmental problems would exist. "The whole drama would be played out before empty stalls...." A better understanding of the brain is certain to lead man to a richer comprehension of himself, of his fellow man, of society, and, in fact, of the whole world with its problems, some of which are discussed in this book.

Without any qualification, the human brain is the most complexly organized structure in the universe. The brains of higher mammals appear to be not greatly inferior in structure and in most aspects of operational performance. Yet there is something very special about the human brain. Its performance in relationship to culture, to consciousness, to language, to memory, distinguishes it uniquely from even the most highly developed brains of other animals. But we cannot yet comprehend how these subtle properties came to be associated with a material structure that owes its origin to the biological process of evolution. Our brains give us all our experiences and memories, our imagination, our dreams. Furthermore, it is through our brains that each of us can plan and carry out actions and so achieve expression in the world as, for example, I am doing now in writing this account. For each of us our brain is the material basis of our personal identity—distinguished by our selfhood and our character. In summary, it gives each of us the essential "me." Yet, when all this is said, we are still only at the beginning of comprehending the mystery of our being.

Let us investigate the brain as we would investigate a machine—a special kind of machine of a far higher order of complexity in performance than any machine designed by man, even the most complex of computers. First, we will examine the structure of the brain—the components from which it is built and how they are related to each other. Second, we will look at the way in which its simplest components, the nerve cells (called neurons), operate. Third, we will explore the linkage of individual nerve cells into the simplest levels of organization. We know something of all of these. However, we are still at a very early stage in our attempts to understand the brain, which is the last of all the frontiers of knowledge that man can attempt to pass and encompass. We'll never run out of questions on this greatest of all problems confronting man, because they multiply far faster than we can answer them! Vigorous and exciting new disciplines have emerged in such fields as neurogenesis, neurocommunications, neurochemistry, and neuropharmacology.

NEUROGENESIS

The building of the nervous system, of which the brain is the principal component, provides the most wonderful example of organic growth. As we study, stage by stage, the development of the brain from the earliest embryonic period onward, we are confronted with a dramatic story. In the neural plate, which gives rise to the central nervous system, the primitive cells multiply. Eventually, they form nerve cells that grow fibers which connect with other nerve cells. It is an almost infinitely complex, programed performance, as if there were some supremely intelligent conductor in command. It is believed that this development, down to every minute detail of the fully formed brain, is the result of two kinds of instructions. First are the genetic instructions passed on from the parents and coded in deoxyribonucleic acid (DNA) (see Chapter 9). These instructions determine and guide the cell's synthesis of its many different enzymes. The enzymes are able to increase greatly the rate of chemical reactions within the cell; each enzyme is responsible for a specific type of reaction. Therefore, the synthesis and degradation of the cell's components and, hence, the structure and growth characteristics of each cell depend on its array of enzymes.

Second, the enzymes and other structural components whose synthesis is determined by the DNA represent a source of secondary instructions, which, at all stages of development, guide the growth of neurons by sensing specific chemical signals that are believed to derive from the special protein structures in their surface membranes. Thus, for example, nerve cells in the cortex of the brain develop axonal processes, which, as they pass through a maze of other cells and their complex of intermingling extensions, enter well-defined tracts in the brainstem and continue down the spinal cord for several feet, finally terminating on specific spinal nerve cells. These, in turn, also have axonal processes that grow out from the spinal cord to reach specific muscles. Sensory nerve impulses from the retina of the eye, the cochlear organ in the ear, or receptors in the skin traverse specific nerve-fiber pathways that often are interrupted by several relay stations, through which the impulse is passed on to other nerve cells before arriving at precisely localized portions of the brain. In addition, cells from different portions of

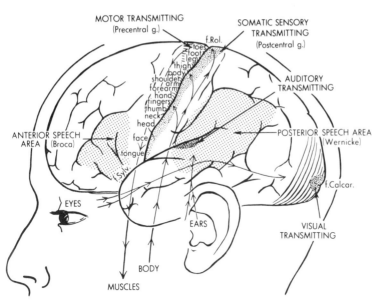

FIGURE 1 *The motor and sensory transmitting areas of the cerebral cortex. The approximate map of the motor transmitting areas to the various areas is shown in the precentral gyrus. The somatic sensory receiving areas are in a similar map in the postcentral gyrus. Other primary sensory areas shown are the visual and auditory, but they are largely in areas that cannot be seen from this lateral view. Also shown are the two speech areas.*

the nervous system are connected in highly diffuse networks. This permits complex integrations among many separate sensory and motor functions.

All of this detailed connectivity is established before any organized activity begins in the embryonic brain!

Research on the building of the nervous system is now being carried out at a high level of scientific expertise. The growth of fragments isolated from embryos is under intensive study. Electron microscopy provides a tool to help evaluate the stages of development of those nerve cells. Various kinds of chemicals, viruses, and radiation damage are used to destroy specific components of the nervous system, and the reactions of the rest of the system to such damage are being studied.

Powerful techniques of radio-labeling also contribute to the understanding of neurogenesis, particularly in sites that are favorable for rigorous investigation. In these methods, radio-isotopes are incorporated in the nuclei of cells before cell division begins; in this way the cells can be identified and followed for long periods of time by tracing the radio label they carry. Thus we are beginning to understand how defects arise in brain development. From this will come the rational treatment of those congenital defects in the brain that arise from enzymatic or genetic disorders (see Chapter 9). For instance, there are a large number of so-called "inborn errors of metabolism," in which the lack of a particular enzyme may lead to a disorder of the nervous system. Investigations of these disorders have led scientists to a better understanding of *normal* biochemistry, as well as suggesting potential methods of treating the ailments. One notable class of such disorders are the lipodystrophies, which are characterized by an abnormal accumulation in the cells of un-metabolized lipids. Such concentrations of fatty substances disrupt normal functions of the nervous system. (The best known lipodystrophies are Niemann-Pick, Tay-Sachs, and Gaucher's diseases.) In addition, neurological defects caused by such environmental damage as malnutrition might be treated before and after birth, as indicated by Dr. Mayer in Chapter 7.

NEUROPHYSIOLOGY

Investigations on the functional performance of the nervous system have been carried out for many decades, but only in the last 20 years have microelectrode techniques been utilized effectively. These are now most important in all investigations concerning the operation of neuronal machinery, both in localized regions of interaction and in the long, complex pathways of operation. Scientists now know most of the various modes of communication between nerve cells across their regions of close contact (called synapses). The structure and function of several regions of the brain, such as the cerebellum, are also well understood. All the neuronal constituents are recognized and their synaptic connections also have been established. However, much more investigation is needed to unravel the organization of those neuronal systems that operate over longer distances and are integrated more complexly.

A most important development for the future is the analysis of the firing frequencies of the individual nerve cells. These are the basic units of the brain, both structurally and functionally, and they total some tens of thousands of millions. In any one zone of the brain there are, of course, multitudes of similar nerve cells, so that observations of single cells are a valid sampling method. Our laboratory is particularly concerned with this study of impulses discharged by single cells. One must regard the nervous system as being designed essentially for conveying and integrating information that is coded in these patterns of impulse discharge.

Although such brain research is based simply on a desire to know more about nature's structure and function, it has made possible many practical applications. For example, electrical recordings from the brain produce the electroencephalogram, which provides information about such brain disorders as epilepsy. This information is of great value in diagnosing the ailments, localizing them in specific areas of the brain, and treating them. Another important application is in electrical recordings from peripheral nerves and muscles, which lead to the diagnosis and understanding of many disorders of movement.

NEUROCOMMUNICATIONS

Neurocommunication is essentially concerned with the transfer and integration of information as it moves through the nervous system. Patterns of impulse frequency, referred to under "Neurophysiology," provide the raw material for the development of neurocommunications theory. However, it must be recognized that the nervous system does not operate simply by single lines of communication; rather, multitudes of lines run in parallel. Superimposed upon this pattern is an amazing complexity of dynamic loop operation, with all manner of feedback controls and cross-linkages. We must envisage principles of design in this network that extract precision of performance from a structure operating with considerable "background noise," much as in some tape recorders and phonograph records. Besides the "noise" there are aberrancies of connectivities that one would expect for a biological system that has grown in the manner outlined under "Neurogenesis."

More and more, attempts are being made to understand the mode of operation of neuronal systems by using a product of man's brain—computer models designed to simulate the performance of portions of the brain and, at the same time, to embody the principles of design that have been discovered in the neuronal system.

NEUROCHEMISTRY

Neurochemistry is a relatively new specialty, which has developed in an amazing manner through the use of many microchemical procedures and radio-tracer techniques. These techniques are so refined that investigations can be carried out even on single cells. By using histochemical fluorescence methods, mapping of the nervous system can be accomplished by tracing the predominant transmitter substance along fiber pathways. Thus, for example, separate neuronal pathways have been identified that contain mainly dopamine, norepinephrine, or serotonin—three transmitter substances that are known to affect the functions of neurons. Sometimes, the determination of these pathways has coincided with pharmacological, physiological, and pathological observations, all of which combined have provided understanding of major nervous-system functions and diseases. An important example of the potential value to mankind in this combined neuroscience approach is the modern-day therapy for Parkinson's disease with

L-dopa, which is discussed in more detail under 'Neurological Disorders" on page 80.

Some other aspects of these combined observations are concerned with the molecular structure of cell membranes and with synaptic receptor sites (see below) and associated channels that transport sodium, potassium, and chloride ions, among others. These ions are the basis of the changes in membrane potentials, which are essentially concerned in neuronal operation.

Other key problems of neurochemistry concern the specific materials manufactured by cells and transmitted at the synapses, and the specific receptor sites for those transmitters. Such transmitters, through their reaction with specialized portions of nerve membranes, permit the passage of an impulse from one nerve to the next. Thus they are part of the molecular basis for the complex communications among nerves in the brain. For example, acetylcholine has been recognized for many years as one of these transmitters. Only recently, however, have scientists been successful in extracting from the specialized receptor membrane the specific molecular component that combines with acetylcholine. This is an important advance toward explaining the actual mechanism by which communication among neuronal structures is achieved and regulated. Other transmitters, such as the amines mentioned above and certain amino acids, may have other types of receptors and may function in different portions of the nervous system.

This branch of the neurosciences will ultimately illuminate both the trophic, or "nourishing," relationships between nerve cells and surface sensing, which many scientists believe is the basis of all the connections that are established between nerve cells during their development. Surface sensing means that the outer membrane of the cell can receive information at certain sites from the surfaces of other nerve cells and react appropriately, either by making contacts or by avoidance.

Neurochemists are also examining many other fields,

specifically enzyme systems and metabolic cycles. The brain has a remarkably high metabolism, which is concerned, to a large extent, with the operation of pumps that actively move such ions as potassium and sodium across the surface membranes of nerve cells. These ionic pumps help to maintain the correct ionic composition of nerve cells and the associated electrical potentials across their surface membranes. Metabolic energy is also required for pushing the trophic substances along the nerve fibers, which, of course, form the lines of communication that are either within the nervous system or in the peripheral nerves.

Despite the high metabolism of the nervous system, the average energy per nerve cell is extremely low. It is the enormous *number* of nerve cells which, collectively, give the brain its high metabolic rate.

NEUROPHARMACOLOGY

Neuropharmacology is closely allied with neurochemistry, and is especially concerned with identifying the action of the various synaptic transmitters and attempting to analyze the way in which they work. In these studies, recent developments of micropharmacology have become increasingly important. One method used in these ultraprecise techniques is to apply specific substances to the exterior of the cell membranes through micropipettes. By this means, several synaptic transmitters of the central nervous system are now known with reasonable assurance: acetylcholine, glycine, gamma-aminobutyric acid, and norepinephrine. These substances are secreted by nerve terminals at the synapses and form the means of transmission from nerve cell to nerve cell across the synapses.

A most important field is investigating the ways in which the brain manufactures, emits, and removes these synaptic transmitters. Essentially, we still operate conceptually with the key-lock model, both of transmitter action at the *receptor sites* on surface membranes of nerve cells and of its simulation or blocking by pharmacological agents. There is an immense future in this field of neural control. One can predict with assurance that there will be great successes in the efforts to develop specific chemical substances that have unique controlling actions upon particular aspects of brain function. The great improvements in anesthetics can be mentioned, and already there are effective controls of epileptic seizures and of the peripheral paralysis of myasthenia gravis.

Of course, grave dangers attend the unrestricted use of substances that affect brain function. This is evidenced by the growth of drug habits. Specific pharmacological actions of such hallucinogens as LSD, for example, can—unless they are administered by knowledgeable doctors—cause serious psychotic disorders and even the permanent destruction of personality. The same dangers apply to other forms of drug addiction, which are now under intense study in laboratories both here and in other countries.

MEMORY

Investigations of memory have developed in two quite separate ways. One line of research is physiological and has a counterpart in neuroanatomy. It is concerned with how use and disuse affect the ability of the synapses to transmit information and to correlate any changes with synaptic growth and breakdown. In the incredibly complex communication system of the brain, each memory must involve some stabilization of specific channels. A sensory input—something seen or heard, for example—would thus be able to trigger a whole series of patterned activites that are much the same as those produced by an earlier input of the same kind. One can say that this replaying in the brain results in remembrance in the mind.

The other line of investigation is concerned with the postulated chemical mechanisms of memory, and seeks to discover specific chemical molecules that have a unique relationship, each with a specific "memory." This work can be criticized because it is not effectively correlated with neuroanatomy and neurophysiology, and provides no basis for recall, which is the essence of memory. Nevertheless, the chemical investigations are important because memory must be the result of some neuronal change that has a chemical and metabolic basis. For example, there is now good evidence that long-term memories cannot be laid down if the enzymes responsible for protein synthesis in the brain are inactivated. Even the RNA responsible for the production of these enzymes is essentially concerned in the laying down of long-term memory traces.

SPEECH AND CONSCIOUSNESS

During the last decade, remarkable studies have been made by Roger Sperry and his associates at the California Institute of Technology. They have worked with patients in whom epilepsy was so uncontrollable that to relieve the seizures it was necessary to cut the great commisure (the corpus callosum) connecting the two cerebral hemispheres. In all the patients, the centers for speech were in the left, or dominant, hemisphere, as is almost always the case. Sperry showed that all conscious experiences of the subjects were derived from events that took place in the neurons of this dominant hemisphere. The minor hemisphere could program a wide variety of skilled sensory responses, but none of them gave conscious experience. Evidently only the speech area and the associated ideational areas of the dominant hemisphere give us our conscious experiences. This remarkable finding is of the greatest importance both neurologically and philosophically.

NEUROLOGICAL DISORDERS

It must be realized that the basic scientific investigations discussed here are essential to the rational treatment of neurological and psychiatric disorders. Obviously, adequate treatment of a malfunctioning machine such as the brain must depend upon an understanding of how it operates. At present, our understanding is very imperfect, yet there have been some remarkable successes. Special mention should be made of the fine work that has been done on substances called catecholamines. These are found in high concentration in certain nuclei of the brain stem and spinal cord, and it has been shown that catecholamines are directly related to our "moods." As a result, a whole new vista has been opened for treating psychiatric disorders that once were believed to be incurable.

A remarkable example of a treatment that derives from chemical and pharmacological studies of the brain is the use of the substance L-dopa for Parkinson's disease. This progressive illness is characterized by muscle tremors and weakness, difficulties in walking and speaking, and slowing of voluntary movements. Parkinsonism afflicts more than a million Americans today, and for many years was treated by destroying certain areas of the basal ganglia, which are masses of gray matter deep in brain. Then it was discovered that two nuclei in this area normally contain a very high content of dopamine, which is believed to function there as a synaptic transmitter. In Parkinsonism the dopamine content is greatly reduced. Dopamine itself could not be given directly to patients because the blood-brain barrier blocked it from reaching neurons in the affected regions. However, it was found that the near chemical relative, L-dopa, *could* cross the barrier and be changed to dopamine when it reached the basal ganglia. Under carefully controlled conditions, its administration greatly improves or checks the symptoms of Parkinsonism, although it does not prevent the progressive neural degeneration.

Another important advance in recent times has been in the study of demyelinating diseases, in which myelin, which forms an insulating wrapping around nerve fibers, is damaged, with the result that impulse transmission is blocked or interrupted. This occurs in multiple sclerosis, for example. The cause of this degeneration is not known, but some investigators attribute it to a virus that

FIGURE 2 *Rat nerve wrapped in myelin membrane. Magnification is X 70,000.*

has lain dormant in the body for many years. Although no cure is now known, one can predict that methods will be devised in the near future for activating cell systems to produce new myelin and wrap it around the nerve fibers, restoring impulse transmission and, hence, relieving the progressive invalidism now caused by the disease.

Necessarily, the treatment of neural and psychiatric disorders has to be carried on despite the inadequacy of our understanding. All levels of investigation must be carried out in parallel. Certainly, the great successes in all the basic scientific investigations will be paralleled by successes in the understanding and treatment of neurological disorders.

CONCLUSION

At the present time, we can think of the total problem of brain operation as a pile of disconnected fragments of a jigsaw puzzle with only a relatively few fragments pieced together to give a meaningful understanding. But every year many more intelligible parts of the picture emerge as pieces of the puzzle are put into position. Special successes occur when various small, organized regions are seen to fit together.

Currently, all of the various fields of investigation outlined above are being pursued effectively in many laboratories throughout the world, but much more effort could be used fruitfully. It is important to recognize that brain research is not a restricted field of scientific investigation, but involves a study of the most complex structure in nature, which is, in addition, the physiological basis of mind. Understanding this structure will require the highest intellectual efforts, not only of experts in the various biological disciplines specified here, but also of theoreticians and experimentalists from all the physical sciences—in particular, mathematicians, chemists, biophysicists, and bioengineers.

CHAPTER 6 SHELDON J. SEGAL

Population Growth: Challenge to Science

Man is losing that balance with nature which is an essential condition of human existence. With that loss has come a loss of harmony with other human beings. The population problem is a concrete symptom of this change, and a fundamental cause of present human conditions.

From *Population and the American Future*
The Report of the Commission on Population
Growth and the American Future, 1972

THE PROBLEM

THE POPULATION OF THE WORLD is now about 3.6 billion and we are adding close to 70 million people a year. At this rate, by the turn of the century the human population will be over 7 billion. It took all of history to the year 1850 to produce a world population of one billion; it took only 100 years for the second billion, and 30 for the third; it is taking only about 15 years for the fourth and it will take less than 10 years for the fifth billion. What these striking figures indicate is that the world cannot sustain such a growth for very long. Either the birth rate must come down or the death rate will go back up.

The current, unprecedented growth rate is caused not by an increased birth rate but by a decline in the death rate. That decline is the result of improved food production and distribution, a more effective social organization in many parts of the world, and particularly the mass application, in recent decades, of such modern public health methods as vaccines, antibiotics, water sanitation, and nutritional programs. The death rate has come down, that is, because significant advances in modern science and technology have been exported successfully from the economically privileged to the less-privileged countries. But factors leading to a reduction in the birth rate have not been transferred so easily from country to country.

THE CONSEQUENCES

The consequence most usually associated with the population problem on a global scale is insufficient food. This is a serious matter, irrespective of the fact that there is a difference of opinion among agricultural experts. Some predict that the world will soon be in bad trouble on this score; others believe that, thanks to the "Green Revolution," we shall be able to avoid the disaster of massive food shortages (see the chapters by Jean Mayer and Herman Mark). The range of informed opinion, from occasional famine to bare subsistence, is not a comfortable one. At best, it appears that the race between numbers of people and amounts of food will be close.

The issue is considerably broader than the difference between a good diet and a subsistence diet. The very hopes for economic development by the poorer countries are at risk. In such countries, a large part of the economic output must be consumed daily by the rapidly growing population in the struggle for immediate survival. These countries cannot easily invest part of the national income for the factories, roads, fertilizer plants, irrigation networks, and machinery that will yield a better life tomorrow. Adding to the burden, high birth rates lead to a large proportion of the population in the dependent years, when they require investment in such social systems as education and health, but are not yet able to contribute to the economy. In many developing countries, more than 40 per cent of the population is under 15 years of age, as against 25 per cent in the developed nations. Thus, the countries that have both a large overall growth rate and a high ratio of dependent youth are those that can least support the burden.

In countries already industrialized, the effect of rapid population growth is expressed in ways that are different, but of no less consequence. Rapid population growth and urban concentration intensify the dilemma of environmental deterioration and heat many social problems to the boiling point. Congested cities, rising crime, and feelings of anonymity cause considerable distress. The demands for enrollment are straining the educational facilities in even the most affluent countries. Primarily, pollution results more directly from *what* people do rather than *how many* people there are; and congestion is the result of uneven population distribution rather than a

This aerial photograph of Coney Island could be a doomsday forecast of the shrinkage of living space.

saturation of the available land area. Nevertheless, these twin problems of modern society—congestion and pollution—become more and more serious threats to the quality of life as the population numbers increase.

Another consequence of high birth rates pertains to people of all countries, rich and poor alike. It relates to the extension of man's control of circumstances—the effective personal freedom of couples to determine the number of children they will have and when they will have them. Throughout the world today, couples do have that freedom in law or in principle, but because of ignorance and poverty, because of lack of proper information and services, the poor people who do not have adequate medical care are not equal in this respect with those who are better off. Thus, the burdens of excessive unwanted fertility, with its tolls of health liabilities to the mother, increase in physical and mental defectiveness among newborn children, and demands on the already depressed economy of the family, fall heavily upon those who are economically deprived.

For instance, in Thailand, Indonesia, and Mexico, the percentage of all births per year by women under 18 years of age or over 35—the age groups with the highest risks to both infants and mothers—is close to 25 per cent. In the United States, Britain, and Sweden, by contrast, the percentage is less than 10 per cent. In terms of medical complications of pregnancy, even in developed countries the under-18 and over-35 groups are more liable to develop diabetes during pregnancy, to become anemic, and to deliver damaged babies; in the developing countries, these problems drain scarce medical resources, as well as causing human tragedies.

THE COURSE OF ACTION

Historically, the transition from a nonindustrialized, poor economy to a modern economy paves the way for the transition from high to low fertility. A high rate of population growth may itself be the most serious obstacle to that transition. As a result, in the coming years, each country will adopt a population policy and seek a means for its successful implementation. The population problem already has taken root in governmental policy. More than a score of countries have established national programs to bring family planning to their people, and the debate is not *whether* to adopt family planning, but *how* to mount effective programs. Means to regulate fertility are provided as an integral part of the health services made available to the 800 million people of China and the half-billion of India. Approximately 90 per cent of the population of Asia live in countries with national family planning programs, along with 25 per cent of the people of Africa and 20 per cent in Latin America. The list includes, in addition to China and India, South Korea and Taiwan, Malaysia, Indonesia and Singapore, Ceylon and Pakistan, Iran and Turkey, Egypt, Tunisia and Morocco, Kenya and Ghana, several of the Caribbean islands, Mexico, Chile, and Colombia.

The results have been mixed. The Indian government took the initiative by launching its official program in 1954. The first decade was rather ineffectual, so results can be measured realistically only since 1964. Even this pioneering operation can boast today only about 12 per cent of India's couples of reproductive age using contraceptives—although this represents more than 12 million people. On the other hand, about 50 per cent of the women in Korea and Taiwan use contraceptives, thus bringing those countries closer to the level of the industrialized nations and so representing a reversal of the traditional pattern in most developing countries.

There are three ingredients for a successful family planning program: interest on the part of the people, effective and acceptable contraceptive technology, and the organization to bring one to the other. Studies by social scientists have shown that the first exists in surprisingly large measure. Throughout the world, substantial proportions of the people want to control family size. The third ingredient, program management, is in the realm of public-health administration, and much improvement is needed. Many national programs still lack an effective, skilled organization to administer and conduct the necessary effort. Yet, this is not the only obstacle. Experience has shown that an issue of critical importance is the contraceptive technology itself. And herein lies a challenge to modern science. There is no definite evidence that new contraceptive methods are needed to achieve wider acceptance and practice of fertility regulation. Yet, the contraceptives people now use have shortcomings and disadvantages. They tend to fall on a very thin line between safety and effectiveness. Modern man arrives home in his luxury autombile, opens his garage door by remote control, enters his climate-controlled home, and one-button tunes his color television set to watch men ride on the moon. Yet, his methods of fertility control are far below the category of a science spectacular. Until recently, all methods of contraception were based on the simple and direct principle of preventing an encounter between sperm and egg. The sperm, as part of the male ejaculate, have been confronted with vulcanized roadblocks or plunged into lethal pools of jellies, creams, or effervescent fluids. These unsophisticated, albeit effective, procedures can hardly be pointed to with pride as achievements of modern science. With a growing understanding of the reproductive process, opportunities for safe, controlled interference have become evident.

THE FACTS OF LIFE

The process of reproduction in all mammals, including man, may be viewed as an intricate series of events which must proceed in perfect succession. Any interruption of the sequence can stop the process. The role of the male ends with fertilization. The female, however, continues with the complex process of harboring the newly fertilized egg in a remarkable nutritional and protective matrix, controlled by a variety of hormones.

The human female, from the time of sexual maturation, prepares approximately each lunar month for the possible occurrence of a pregnancy. The changes that this entails are controlled by a series of chemical messengers (hormones) that pass from their production sites (endocrine glands) via the bloodstream to specific parts of the reproductive system. Each month two different sequences of events are repeated. From among the tens of thousands of immature eggs in the ovaries, several will start to develop, but after about ten days only one will continue to flourish and become fully mature, ready to be released at about mid-cycle on approximately the fourteenth day. Ovulation occurs and the released egg is swept from the surface of the ovary by the undulating open end of the fallopian tube.

Fertilization may occur if, around the time of ovulation, sperm ascend to the fallopian tube. At this time the reduction of the egg's genetic material by half is completed, so that fusion of the sperm nucleus and the egg nucleus produces the normal complement of hereditary factors. Next, a slow series of cell divisions begins. After thirty-six hours, the single cell has become two. Two days later the fertilized egg may have divided two more times to form a microscopic ball of eight cells. In this condition, the egg will complete its descent of the fallopian tube and pass into the uterus.

Four days after fertilization, the egg is a cluster of 32 or 64 cells, beginning to divide more rapidly. This corresponds to about day 19 or 20 of the menstrual cycle. Very little further development can occur if the egg remains free within the uterus. It may float unattached for one or two days and assume the form of a signet ring, with an inner mass of cells encircled by a single row of cells in alignment. This pre-embryo state is called the blastocyst. Under proper conditions, the ring of cells will nestle into the uterine wall and begin to form the placenta. The inner cell mass, after several more days of cell divisions and internal rearrangements, will become a human embryo. This is the history of an egg.

Meanwhile, a second sequence is taking place to assure a safe and supportive nesting place in the uterus.

Early in the menstrual month, before ovulation, the ovary secretes in ever-increasing amounts the female sex hormone, estrogen, which enters the bloodstream and reaches the uterus. It stimulates the lining of the uterus, called the endometrium, to proliferate, increasing its thickness perhaps fourfold, and to become much more vascular.

At about the time that ovulation is occurring, the ovary begins to produce a second hormone, progesterone, and the production of estrogen falls off considerably. The cells of the endometrium become more corpulent. The endometrial glands grow rapidly in length and thickness and begin to accumulate secretions. By day 20 of the cycle, the entire inner surface of the uterus has become a highly vascular, spongy bed ready to accept, protect, and nurture a fertilized and dividing egg, if one arrives from the fallopian tube.

The uterine lining is now under the remarkable influence of progesterone, and if a pregnancy is to be established and an embryo formed, the progesterone support must continue throughout the remaining 35 weeks of pregnancy. Without it, the blastocyst, or new embryo, would pass out with the sloughed-off endometrium and the menstrual blood at the time of the first expected menses.

The synchrony of these two sets of events—the development and fertilization of the egg on the one hand, and the preparation of the uterus as a proper environment for the attachment of the fertilized egg on the other—is the result of the highly integrated functions of the hormones involved in each process. For example, the same pituitary hormone that stimulates the ovary to begin the maturation of an egg stimulates other cells of the ovary to produce estrogen. The hormones that maintain the uterus in a receptive state throughout pregnancy prevent further cycles of egg development and release, thus assuring against the occurrence of ill-timed, superimposed pregnancies.

With the advent of "the pill," contraception caught up with the twentieth century. The method is based on preventing egg development and release, using hormones similar to those that prevent these events during a normal pregnancy. The discovery of "the pill" required the participation of chemists, biologists, and physicians. It was the product of countless experiments by scientists of many countries. Yet, it is only one means to interfere with but one of the vulnerable links in the reproductive chain of events. The discoveries of basic science define a substantial number of potentially vulnerable links that could provide advantages over ovulation suppression as a means to achieve effective and safe contraception. The list of such links includes egg and sperm transport to the fallopian tube, the fertilization process itself, early development of the fertilized egg, preparation of the endometrium, normal synchrony of the necessary hormones, and, in the male, sperm production, maturation, and transport. Presently, scientists are probing each of these events, in efforts to bring about controlled interference in a manner that would result in effective and safe contraception. Envisaged as possible in the not-distant future are contraceptive methods that would take the form of daily, weekly, or monthly pills for women or men, yearly inoculations for either men or women, or subdermal implants of small plastic tubes that could release fertility-controlling hormones for several years. This method, too, is being developed for possible use by either sex.

More research, by far, is being done to study the female reproductive system than that of the male. Accusations by feminists notwithstanding, this concentration of effort on topics most likely to lead to methods for use by women is not an expression of male chauvinism by male researchers and male policy makers. It simply reflects the fact that the female system offers more points, by far, for controlled interference than does the male system. This is shown in the following diagram.

The list of vulnerable links grows with each new dis-

Vulnerable steps in the reproductive process, female and male

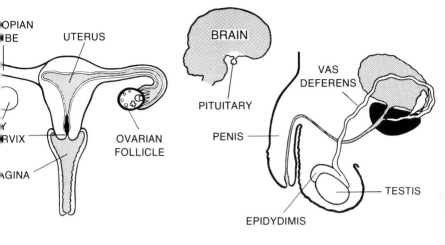

FEMALE

Ovulation

Egg production
Egg maturation in ovary
Egg release (ovulation)

Ovum Transport

Transport of egg to fallopian tube
Egg transport in fallopian tube
Fluid secretion by fallopian tube

Sperm Transport

Sperm transport through cervix
Sperm transport to arena of
*   fertilization (in fallopian tube)*
Sperm movement to egg surface

Fertilization

Sperm-egg interaction (beginning of
*   fertilization)*
Fusion of egg and sperm nuclei
*   (completion of fertilization)*

Transport and Implantation

Transport of fertilized egg along
*   fallopian tube*
Early development of fertilized egg
*   (blastocyst development)*
Preparation of endometrium for
*   implantation of blastocyst*

Development

Placenta formation
Maintenance of pregnancy

MALE

Sperm formation in testis
Sperm maturation in
*   epidydimis*
Sperm transport in vas
Seminal fluid biochemistry

covery in the realm of basic research. Scientists have learned recently that sperm produced and released from the testis are still immature and require a storage period of about 30 days to undergo maturation. This process, surprisingly, can be prevented by the use of anti-hormones that may prove to be otherwise innocuous. Even more unexpected was the discovery that the maturation of sperm continues for about six to 12 hours, even after they are deposited in the female and before fertilization, thus providing another link for controlled interference.

THE CHALLENGE

These are but a few examples of a constantly growing body of knowledge concerning the long-neglected subject of human reproduction. The challenge to modern science is to broaden even further the base of understanding and to find ways to translate these findings into technological advances for the control of fertility. The world's diverse population, with its myriad cultural, religious, nutritional, and constitutional differences, demands a number of methods to choose from if wide-scale success in the regulation of fertility is to be achieved. The achievement of this goal could very well be the most important contribution the life sciences have made to mankind.

CHAPTER 7 JEAN MAYER

Nutrition and Research

> *The nutritional deficiencies retarding the health and performance of large numbers of the world's people are probably the most urgent resource problem we face today.*
>
> From *Natural Resources*. A Summary Report, Committee on Natural Resources, National Academy of Sciences-National Research Council, 1962

THE SCIENCE OF NUTRITION is very young. Even though food has been man's main preoccupation since the beginning of his existence as a species, nutrition as a science is less than two hundred years old. Casimir Funk, who coined the household word vitamin, died in 1967. By contrast, while the stars are far away and exert, at most, an extremely indirect influence on the human species, astronomy became a science more than 25 centuries ago. But then, experimental biology—in particular, modern physiology—started only at the beginning of the seventeenth century in England, when William Harvey discovered the general circulation of blood.

Very broadly, the evolution of the science of nutrition can be divided into four periods:

1. the prescientific period, extending from the dim beginnings of the Stone Age to the second half of the eighteenth century, after Antoine Lavoisier, the great French chemist (1743–1794), introduced his heat-measuring concepts;
2. the age when caloric and nitrogen balance were studied in the nineteenth century;

3. the era when the trace elements, vitamins, and essential amino acids were discovered and the deficiency diseases studied, extending from the beginning of the twentieth century to the 1940s; and
4. the period since the end of World War II, with development of the study of the role played by nutrition in degenerative diseases, and the recognition that what is in the diet can be almost as important as what is missing from it.

Let us review briefly the history of these periods.

THE PRE-EXPERIMENTAL ERA

An intricate knowledge of food is necessary for survival. In a long series of trials and errors—sometimes mortal errors—the number of animals and plants known to provide food was slowly extended. And the idea that food does more than assuage hunger is very old. Certain beliefs, such as the notion that consumption of the heart or flesh of brave or strong animals or enemies would confer those same virtues on the warriors of the tribe, appeared during the Stone Age. While some edicts may well have had something to do with the unwholesomeness of certain foods (e.g., the biblical prohibition of pork may well have been based on the recognition that illness, and even death, could result from eating swine now assumed to have been infested with *Trichinella*), most such beliefs seem to have had no basis in fact. Examples of apparently pointless practices were the Egyptian prohibition of beef or chicken for kings and the biblical declaration against hare.

Closely allied with the problem of taboos and superstitious beliefs was the search for foodstuffs that also could serve as remedies. Many herbs and parts of plants or animals were prescribed because of their shape, their color, or some other property unrelated to any demonstrable pharmacological effect (Figure 1). But the common recommendation of the "Ebers Papyrus" (1600 B.C.); of Hippocrates, the Greek physician who was born

FIGURE 1 *The mandrake plant* (Mandragora officinarum), *whose roots often resemble human limbs, has, since earliest antiquity, been used medicinally and as a magic potion. It has been used as a love potion, as a "cure" for many ills, and supposedly promoted fertility in women.*

about 460 B.C. and was the founder of scientific medicine; and of several medieval writers to the effect that liver was a "cure" for eye diseases and for night blindness; and of Cartier's Indians, who used infusions of evergreen needles for scurvy, are early examples of prescriptions for deficiency diseases. Incidentally, Hippocrates paid considerable attention to nutrition in his writings. Many of his opinions are difficult to justify—for example, his belief that beef is more troublesome to digest than pork, or that for feverish patients fish should be roasted rather than boiled. But his abhorrence both of extreme abstemiousness and diet restriction and of excessive food intake without corresponding physical labor are valid nutritionally.

ENERGY AND PROTEIN

Probably the greatest turning point in the history of biology and medicine was the investigation of combustion, conducted during the 1770s by the French chemist Antoine Laurent Lavoisier. He introduced measurements into biological and chemical studies, and his experiments laid the basis for our understanding of caloric expenditures and requirements (Figure 2).

Specifically, Lavoisier showed that the body uses oxygen to burn foods and that the smallest amount of oxygen is used when the individual is resting at a comfortable temperature, several hours after he has eaten. This is very close to what we call the condition of basal metabolism. The figures Lavoisier obtained for oxygen consumption, using primitive equipment, were near to the figures we now recognize as normal. Lavoisier showed, further, that consumption of food increases the oxygen consumption (and the heat loss) of his subjects, a phenomenon we now call "specific dynamic action." He also demonstrated that exposure to cold similarly increases

FIGURE 2 *Lavoisier in his laboratory during his studies on respiration (from a sketch made by Madame Lavoisier.) This brilliant chemist measured the volumes of air breathed by a man at rest and at work and showed that his temperature remained constant. He concluded "respiration is merely a slow combustion of carbon and hydrogen which is similar in every respect to that which occurs in a lighted lamp or candle....*

*Cochrane, "Lavoisier."

heat loss, energy requirements, and oxygen consumption. Finally, his experiments led to the realization that increased muscular activity increases both oxygen consumption and energy requirements. In fact, such increased activity can double or even triple Lavoisier's basal energy requirement.

Since Lavoisier's studies, a great deal of work has been done to get more precise figures for the energy cost of man's various activities and of keeping body temperature constant under different climatic conditions. As an upshot of this work, we can ascribe a figure, in terms of calories, to all such human needs. We can calculate with great accuracy how many calories a particular man, woman, or child will require to live a particular type of life.

The findings of Lavoisier had to do with *amounts* of food; they told us how much an individual required. They did not reveal whether this food had to be of any particular nature. The next development in nutritional knowledge assured us that it did and, in a general way, indicated the required nature of the food. This took place in the nineteenth century, with the discovery that all forms of calories are not equivalent. Protein, which provides the basis for the construction of the structural elements of the body, plays a very special role. Carbohydrates can replace fat for caloric intake and, to a large extent, fat can replace carbohydrates. For protein, however, there can be no replacements—the body must have a minimum amount if it is to operate efficiently.

Lavoisier's findings were extended by the work of nineteenth-century physiologists and chemists. Scientists in France, the Netherlands, and, pre-eminently, J. von Liebig (1803–1873) in Germany developed concepts and methods of analysis that permitted the establishment of food-composition tables. These, as will be shown, are still the essential tools of the nutritionist, because any reassessment of the value of a diet and any dietary recommendations must be translated in terms of food, if they are to be of practical use.

Jean Baptiste Boussingault (1820–1887), in France, studied absorption and digestion of foodstuffs. The proportions of carbohydrates, proteins, and fats, and their contributions to the caloric content of foods became the basis of nutrition. However, only scant attention was given to the minerals and almost none to the nutritional factors present in small amounts, which, as vitamins, essential amino acids, and essential fatty acids, we now consider to be indispensable to the maintenance of life, let alone of health.

In this country. W. O. Atwater tried to put all practical nutrition on the basis of the financial cost of calories and proteins. In effect, he advocated deriving the diet chiefly from cereals, peas, and beans because of their cheapness, and omitting fruits and garden vegetables, which, he felt—ignoring the detailed nature of diet—contributed little for their cost. Both Atwater and the German physiologist Karl von Voit recommended large amounts of protein for adults (approximately 150 grams per day), in spite of experiments at Yale which showed that men could live and work with less than one-third of this amount.

TRACE ELEMENTS, VITAMINS, AMINO ACIDS, AND DEFICIENCY DISEASES

The revolution in the thinking on nutrition that took place between 1905 and 1910 has rarely been paralleled. It was much like the change after Louis Pasteur discovered that microorganisms can cause infectious disease. The turning point came with the concept of essential nutrients—that is, substances necessary for growth, health, and the maintenance of life. This was tantamount to recognizing that the human organism is a good, but not a perfect, chemist. It can manufacture thousands of complicated molecules. It cannot synthesize a number of structures—vitamins and essential amino and fatty acids. Sir Frederic Gowland Hopkins of Cambridge (England) showed that tryptophan, one of some twenty-two amino acids—the building blocks of proteins—is one such indispensable nutrient. The Americans Osborn and Mendel demonstrated that the addition of tryptophan and lysine, another essential amino acid, considerably improved the biological value of corn proteins; the chief corn protein, zein, is particularly low in these amino acids. Other scientists showed that the minimum adequate diet for the rat must provide, in addition to the long-known nutrients, two unidentified factors, which they called "fat-soluble A" and "water-soluble B." Later, these were shown to be complex: the fat-soluble factors included vitamins A, D, E, and K; the water-soluble factors included vitamin C and several B vitamins.

It is remarkable that although physicians—James Lind in Edinburgh (who recommended citrus fruits for the British Navy to prevent scurvy, which started the nickname "Limey" for Englishmen), Trousseau in France, and Eijkman in Holland—first established nutrition as a factor in such diseases as rickets and beriberi, it is to chemists that we owe the concept of deficiency diseases as clinical entities. It seems almost incredible that barely a century ago Charles Caldwell, one of the most prominent physicians and surgeons of his time, wrote a 95-page pamphlet denouncing von Liebig and proclaiming that chemistry had nothing to contribute to medical science!

It should be unnecessary to point out the extreme speed at which our understanding of the deficiency diseases, the characterization and synthesis of vitamins, and the translation of experimental nutrition into clinical advancement progressed between World Wars I and II. Rickets is a disease of infancy and childhood that causes malformation of bones. Once the most prevalent disease condition in cities of the Western world, rickets was almost wiped out in 10 years by the widespread use of fish-liver oils and by the fortification of milk with vitamin D. When iodine was added to salt, goiter was eliminated from entire populations. This deficiency disease causes the thyroid gland to enlarge, and can be recognized by a large swelling on the neck. Flour enrichment and diet improvement have dealt effectively with pellagra, which manifests itself by abnormal pigmentation and other abnormalities of skin texture and color. In fact, for several years it has been impossible to locate a pellagrous patient in the whole state of Georgia to demonstrate to the students of the medical college of the State University the once-common signs of the disease.

NUTRITION AND DEGENERATIVE DISEASES

Renewed interest in undernutrition during World War II led to the creation of several international organizations and resulted directly in an interest in overnutrition. Observers were struck by the drastic decrease in the number of patients hospitalized with conditions of the heart and blood vessels during the famine that accompanied the siege of Leningrad. When the siege was broken and the famine relieved, the "refeeding" period was associated with an upsurge in such conditions. At the same time, statistics accumulated by life insurance companies in the United States emphasized, although less dramatically, the positive correlation of excess weight and mortality, not only from cardiovascular diseases, but also from liver conditions, diabetes, increased operative risk, accidents, and other circumstances. This body of data, in turn, stimulated interest in the study of the causes and development of obesity and its associative and causative links to disease.

In particular, work conducted in the Nutrition Department at Harvard and, more recently, throughout the world, deepened our understanding of the mechanism of regulation of food intake by demonstrating the paramount role of certain brain (hypothalamic) centers, their anatomical and physiological relationship, and the various ways in which the mechanism could become inadequate, leading to obesity or loss of appetite (anorexia). It was established that, at low levels of physical activity, the minimum voluntary food intake is greater than expenditures, with obesity the result. Certain inherited body types are particularly susceptible to this condition. Obesity is thus seen, in most cases, as a consequence of a mode of life for which many individuals are not genetically adapted.

Considerable interest also was generated by the suggestion that the presence of a high proportion of fat in the diet was associated with a high level of blood cholesterol, atherosclerosis, and coronary disease. A lively and constructive experimental controversy still rages as to the relative influence on cholesterol levels of animal and vegetable fats, the degree of saturation of the fat, the concentration of polyunsaturated (essential) fatty acids, and the effect of plant sterols (wax-like substances analagous to cholesterol, found in man and other animals) on cholesterol absorption. The bulk of the evidence shows that the higher the serum cholesterol level, the higher the risk of cardiovascular catastrophe. Furthermore, it is now agreed that, in general, polyunsaturated fatty acids decrease, and saturated fatty acids increase, the cholesterol level. Putting coronary patients on a "high polyunsaturated" diet—and at the same time reducing the total dietary fat—drastically reduces the risk of a second cardiovascular accident. The favorable effects of placing high-risk individuals on such "prudent" diets have established the scientific background for demands for a change in the food habits of high-risk individuals. In addition, the good results have increased pressures for experimentation in modern food technology in order to change the composition of many of the common (processed) foods toward lower contents of saturated fatty acids, cholesterol, and sucrose.

NUTRITION AND SOCIAL CONSCIENCE

As our experimental and clinical knowledge improved, our sense of social responsibility increased. It came to be accepted that educated nations could not tolerate a situation in which easily preventable diseases continue to kill and disable millions of human beings. In his famous book, *Diet, Health, and Income*, John Boyd Orr (later Lord Boyd-Orr) showed that in Scotland poor growth and sickness were much more prevalent among the low-income classes than among the wealthy. Hazel Stiebling in this country and André Mayer in France conducted nutritional surveys which demonstrated that many children and adults, particularly in the poorer classes, were undernourished or malnourished. In the 1930s, the League of Nations called together a committee of physiologists who promulgated the first set of recommended dietary allowances, as well as a practical handbook on the assessment of the nutritional status of populations.

This impetus was further accelerated by preoccupation with food problems during World War II: scientific advisory bodies, such as the United States National Research Council Food and Nutrition Board, came into existence on several continents. More importantly, more than 60 nations combined their efforts to improve nutrition all over the world. A number of additional nations have joined the organization since, bringing the total to the vicinity of 100. The Food and Agriculture Organization of the United Nations (FAO), the Nutrition Section of the World Health Organization (WHO), and the United Nations International Children's Emergency Fund (UNICEF) are some of the healthy off-spring of this great movement. In many regions, the emergency measures taken by some of the "crisis committees" created by the FAO averted the widespread starvation that would have followed the end of the war. UNICEF has started and encouraged child-feeding programs that benefit millions of children in Asia, Africa, and Latin America. FAO and WHO have initiated, supported, and publicized epidemiological and clinical studies of kwashiorkor—the most widespread deficiency syndrome found in poor areas—which apparently is caused by a lack of high-quality proteins in the diet during early childhood after weaning (Figure 3).

The finding that poor nutrition during pregnancy may lead to an increased frequency of prematurity, with its consequent physiological as well as psychological consequences, is stimulating much-needed research in this

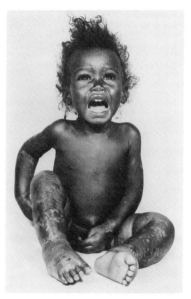

FIGURE 3 *This two-year-old child of Uganda is suffering from kwashiorkor, a protein deficiency disease. Kwashiorkor is a Ghanian word that means "displaced infant." It refers to the fact that the illness often attacks children immediately after they have been weaned and the mother is again pregnant. Symptoms are misery, abnormal swelling, wasted muscles (with fat present), and retarded growth. Note also the "flaky paint" rash.*

area. Similarly, the indication that malnutrition in infancy may also cause mental retardation could well become a powerful goad for the improvement of nutrition of small children in many countries of the world.

In the United States after World War II, the recognition was arrived at very slowly that, although the wealth of the nation as a whole increased, a substantial number of our population continued to be so poor that they could not feed themselves properly. The "hunger movement" took shape only in the sixties as an offspring of the civil-rights movement, spearheaded by Martin Luther King, Jr., and the Southern Christian Leadership Conference. In 1969, the National Council on Hunger and Malnutrition in the United States served as an organizational link among various concerned groups, such as the National Association for the Advancement of Colored People, the National Convention of Negro Women, various church organizations, the United Automobile Workers, and others. Individuals, including the late Senator Robert Kennedy, Senators Joseph Clark, Charles Percy, Edward Brooke and Edward Kennedy, Leslie Dunbar, Robert Choate, John Kramer, Patrick Boone, and Stanley Gershoff were particularly active in waging the war against hunger. "Report on Hunger in America," a CBS television program on the subject, and the newspaper articles of Nathan (Nick) Kotz shook the conscience of the American people and, in 1969, President Nixon called the White House Conference, which achieved enormous publicity and was, in many ways, a watershed in American social history. As a result of its recommendations, 12 million Americans were added to the list of beneficiaries of family food-assistance programs, which were also extended and liberalized. The number of children receiving free school lunches was tripled, free school breakfasts and summer feeding programs were instituted, and a large number of measures related to consumer problems were taken.

In this country, the underprivileged are not the only persons suffering from malnutrition. Much is self-inflicted by middle- and upper-income groups through either ignorance or faddism. This manifests itself in excessive intake of calories—mostly fats and sugar—and in the consumption of saturated fat and cholesterol, which leads to obesity and atherosclerosis. And then there is the other kind of malnutrition that is the result of periodic dieting—usually by teenage girls and young women—which can cause anemia, as well as other physiological damage.

CURRENT RESEARCH

Nutrition research, stimulated by public interest and by progress in understanding the importance of diet in the developmental mechanisms of some of the common diseases, is proceeding vigorously and on a broad front.

We are continuing to identify new required nutrients. For example, fluorine, which we knew was needed in minute amounts as fluoride to protect dental health, has recently been identified as a requirement for mammalian growth, as well. Chromium seems to be needed for several metabolic processes. And a number of other trace elements are under active investigation because research has found that deficiency in these nutrients causes diseases or cessation of growth. The role played by components of the diet in determining the levels of various blood lipids (and individual reaction to these components) is also the subject of particularly active research. The exploration of both cerebral and extracerebral components of the regulation of food intake—the determination of the modes of development of various types of obesity and their relation to the major degenerative diseases—is also one of the primary research areas. The relationship of malnutrition and mental retardation is another subject of current interest now being investigated in a number of countries.

In the technological field, active efforts are being pursued in the development of new sources of high-quality protein—isolated vegetable proteins, protein of microbiological origin, entirely synthetic amino acids—and of inexpensive, high-protein foods, particularly for the feeding of young children. Engineering and modified foods emphasizing long-term medical advantages are also being developed actively. These include replacing meat with vegetable substitutes, which tend to be much lower in fats and free of cholesterol, and foods with low sugar or low salt content. Such developments are also designed to guarantee that more and more people can be fed on produce from less and less arable land, thanks to the revolution in food technology—the so-called "Green Revolution." In Chapter 10, Dr. Mark discusses ways in which by-products of the oil industry have made life easier for man. The new technology might also use these by-products as raw material for food.

Finally, the economic, social, and educational aspects of nutrition as it affects the poor in industrialized countries and the general population in developing countries are at last being given the attention they deserve and require if the physiological problems caused by poverty are going to be remedied. In this connection, a strong argument for controlling the population growth, as described in the previous Chapter, could be based not alone on population problems of the poor, but on those of nations that are becoming wealthier. As people become richer they occupy more space, disturb the environment more drastically, create more pollution, and eat more food than do poor people. The United States is an example. Our population increases, but our main nutritional problem is overweight and our main agricultural problem was, until recently, overproduction!

No science has succeeded in bridging theoretical research and its application to the betterment of mankind as well as has nutrition. At a time when many young people turn away from science because of its lack of "relevance," the science of nutrition stands out like a shining beacon as an example of the fact that while science alone cannot solve the basic problems of mankind, they cannot be solved *without* science. Compassion is not enough, social organization is not enough. Social organization, compassion, and research brought to bear together on the problems of mankind are the winning combination.

HEALTH

During everyone's lifetime, a primary concern is his health—concern that may be as old as language. When we meet a friend, our first question, no matter what country we live in, is likely to be "How are you?" "Comment ça va?" "Wie geht's?" "¿Como esta usted?" Today a great body of scientific research is related to two broad areas of health and disease that can be classified as immunology and genetics. These two subjects embrace health concerns that range from the common cold to cancer, organ transplants, and mental retardation. That is, research scientists are discovering ways to apply insights gained from new techniques, new machines, new concepts. Perhaps just as important, today they are able to know what questions should be asked to derive answers that may one day relieve man of some of his most onerous worries about the well-being of himself and his children. It has been said that a nation's health is more important than its wealth. Our authors tell why.

CHAPTER 8 G.J.V.NOSSAL

The Body's Immune Defense System in Health and Disease

> *Unfortunately, there are enough people who are so infatuated with their specialized studies that they are ignorant and unaware of other disciplines. If fate happens to lead them to fields other than their own, they are helpless and lost....Your country demands more than one-sided proficiency. Only the correlation of extensive knowledge will bring us closer to acutal wisdom.*
>
> Thomas Bartholin (1616-80), in *De theologiae et medicinae affinitate* (tr. by Max Samter)

FEW ADVANCES in preventive medicine have been as dramatic in their impact on human well-being as immunization procedures. Today, we take antipolio drops or antitetanus shots very much for granted. It is easy to forget that a scant hundred years ago the specter of epidemic infectious disease was still a major threat. Yet, the very fact that immunization programs are so widespread suggests at first sight that the science of immunology has passed from the exciting, turbulent research and development stage into a staid, technological phase. In fact, the reverse is true; immunology is one of the most dynamic and expanding areas of biomedical research. Why is this so?

There are three factors behind the current immunology research explosion. The first is the recognition that certain white blood cells called lymphocytes, which are responsible for the body's immune defense against viruses and bacteria, are also crucially involved in a variety of noninfectious diseases. For example, they play a major role in cancer, in the rejection of surgically transplanted organs, and in certain severe internal diseases, such as arthritis, anemia, and nephritis. This has led to the hope that a better understanding of lymphocyte function will eventually provide treatment triumphs of major dimensions.

Second, a dazzling variety of sophisticated new technological tools have become available recently to help in elucidating certain immunological problems. These include electron microscopes, protein-sequencing machines, microchemical and isotopic techniques, and X-ray crystallography.

The third factor stems more from basic science. Immunology has become a meeting ground of unusual value for scientists from various disciplines (such as protein chemistry, molecular biology, cell physiology, and genetics) where frontier problems of vital interest to all parties can be discussed and attacked experimentally. Thus, the immune response can be regarded as a model of a complex set of life processes, involving subtle cell interactions and gene regulations, in which key problems are being unraveled with surprising speed. Solutions gained from immunological studies might yield rules that can be applied in a more general way. Before we can come to a consideration of these frontier issues, it will be necessary to develop some historical perspective.

THE DISCOVERY OF ANTIBODIES

Although it had been known for more than 2,000 years that an attack of a particular infectious disease protects the body against a later recurrence, vaccination was not discovered until 1798. A general practitioner in England, Edward Jenner, followed up a local old wives' tale and showed that the fluid from a cowpox sore could be used to cause a small blister in the skin of a person. The result was protection against smallpox—the related, but immensely more serious, disease. Jenner's procedure was completely empirical—that is, it was based only on experience and observation, rather than on controlled research. The next major step forward did not come until 1880. The great French scientist Louis Pasteur had discovered the microbial nature of infectious disease. He showed that when bacteria were maintained in artificial growth media for many generations, they frequently lost their virulence. These so-called *attenuated* bacteria could act as specific vaccines in experimental animals, and Pasteur used the finding to make a crude (but apparently effective) vaccine against human rabies. In 1890, the German bacteriologist Emil von Behring showed that the serum of animals that had received a vaccine injection contained substances termed *antibodies,* which neutralized the toxic properties of the bacteria or bacterial products that had been injected originally. This neutralizing action was specific; antibodies raised against one bacterial species did not affect other virulent bacteria. The discovery of these basic principles allowed the development of a wide variety of vaccines for human use.

It was soon found that antibodies could be generated against foreign substances other than bacterial products. In a brilliant series of studies spanning the first 40 years of the twentieth century, the Austrian scientist Karl Landsteiner showed that antibodies of exquisite specificity could be produced by injecting into animals harmless substances ranging from foreign red blood cells to egg white and even to such small chemical molecules as dinitrophenol, which had been synthesized in a laboratory and did not occur in nature. It seemed that the injected foreign substance, or *antigen,* caused the production of a strictly complementary antibody molecule, perfectly tailored to fit the antigen, as a key fits a lock.

In the 1930s, it became clear that antibodies were proteins. Proteins are the most important molecules in living matter. They constitute roughly half of the body's materials and include hormones, enzymes, and antibodies, to name only a few. Proteins are important also because they display the greatest diversity in detailed shape and structure, and therefore allow amazingly intricate interaction, such as enzyme action, to take place. Because proteins are long strings of smaller building blocks called amino acids, Dr. Felix Haurowitz, of the University of Indiana, constructed a very reasonable hypothesis to account for antibody production. He theorized that the antigen entered a cell, and there acted as a template, or pattern, around which the long protein string coiled and folded itself, finally assuming a shape imposed by the antigen. That is, the antibody formed in such a way that it could offset or neutralize the antigen.

THE MOLECULAR BIOLOGY REVOLUTION AND THE ANTIBODY PARADOX

According to Haurowitz's "direct template" hypothesis, there was no limit to the number of antibodies a cell or an animal could manufacture; for each antigen, it churned out just the right antibody. The theory was accepted virtually without question for more than 20 years. However, during the decade after 1953, the revolution in molecular biology soon posed major problems for antibody theorists. In England, Drs. J.D. Watson, F.H. Crick, and M.H.F. Wilkins discovered the double-helical structure of the genetic material better known as DNA. (This is discussed further in Dr. Hogness's chapter.)

Overwhelming evidence soon built up that proteins are synthesized according to the dictates of coded instructions in the DNA. For every protein a cell can make, there is a section of the long, stringlike DNA molecule that is called the structural gene for that protein. The section contains a code specifying the particular amino acids that must go into the protein. There are 20 different amino acids, and most proteins contain 100 or more of these units. Moreover, it is not just the overall choice of amino acids that determines the final shape and function of a protein, but more particularly their sequential arrangement in the protein chain. How do we reconcile this primacy of DNA in determining protein structure with the direct template theory? The only way would be if all antibodies contained a *common* amino-acid sequence, but assumed different final shapes. It is now clear that different antibodies have different amino-acid sequences, and also that proteins with a particular sequence cannot do other than to assume a single particular shape, or at most a very few shapes, under physiological conditions of temperature and acidity. In other words, the direct template hypothesis is not tenable.

This confronts us with a major dilemma. The variety of antibody patterns is seemingly endless. The number of genes in the nucleus of a cell, although large, clearly is not infinite. Furthermore, antibody production represents only a small part of the incredible total task that confronts the genetic machinery and that includes everything from the color of our eyes to, possibly, certain behavioral traits. Is there a gene for every antibody an animal can make? How, then, was it possible for evolution so to program genes that antibodies could be produced against synthetic chemicals that had never before existed in nature? We shall see how modern research into the structure of antibodies and the nature of lymphocytes has given us solutions to this puzzle.

CLONAL SELECTION IN ANTIBODY FORMATION

In 1955, the Danish-born scientist Niels K. Jerne, now Director of the Basel Institute for Immunology, put forward the then-unorthodox notion that every antibody which an animal could ever manufacture, to no matter what antigen, was already being made in the cells even before the antigen came along. Naturally, the *amount* of any one type of such "natural" antibody was infinitesimally small. Nevertheless, one milliliter (about 0.0338 of a fluid ounce) of serum contains 10^{17} molecules of antibody (add 17 zeros after 1 to get an idea of the magnitude). If there were 10^{11} different types of antibody, Jerne reasoned, there would still be 10^6 molecules of *each* type in the serum. The role of the antigen was simply to *accelerate by a large factor* the rate of synthesis of the particular antibody that happened to fit it best. In that way, the number of molecules per milliliter of serum might rise from 10^6 to 10^{12} or even 10^{14}, so that now the antibody level might be high enough to neutralize the antigen and so be readily detectable by simple laboratory techniques.

In 1957, the Australian biologist Sir Macfarlane Burnet took this theory one step further. He postulated that the unit of selection on which an antigen works is not the antibody molecule, but the individual lymphocyte cell. He called this his "clonal selection" theory (Figure 1). This says that, although all lymphocyte cells look alike under the microscope, each one is different from other lymphocytes in the way it functions. Each cell has on its surface one unique type of antibody. When antigen enters the body, all it has to do is to find the "right" cell—one with a surface receptor that fits it—and then the cell will go on to divide and simultaneously to build a factory of cellular material for much larger-scale protein synthesis (Figure 2). In other words, the antigen "selects" the lymphocyte to be stimulated, which then produces a "clone"—a brood of progeny cells all derived from a single ancestor. (The word clone is from a Greek word that means "throng.")

Despite Burnet's high status in the world of science, this theory was greeted with shocked incredulity by most scientists, particularly in the United States. In that country, immunology was dominated by immunochemists, who could not accept the idea that such a "sloppy" mechanism would confer such exquisite specificity. After all, they argued, a mouse has only 10^9 (one billion) lymphocytes in its body. Including all the synthetic chemicals that can act as antigens, there must be vastly more than a billion antigens in the universe, each capable of stimulating the mouse to form antibody. The theory must be nonsense!

At that time I was Burnet's student, and it became my task to provide some evidence for (or against!) the outrageous theory. It seemed to me that the theory could be destroyed readily if each antibody-forming cell could make several types of antibody simultaneously. The trouble was that no method existed capable of detecting

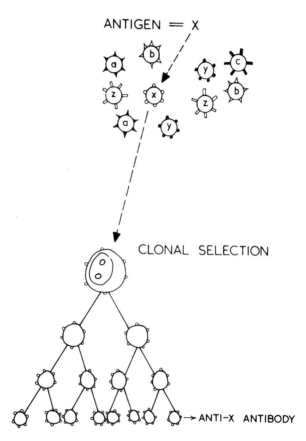

FIGURE 1   *The clonal selection theory of antibody formation. An antigen, X, enters the body. It encounters many lymphocytes, each bearing a particular kind of antibody or immunoglobulin on its surface as a receptor. When the antigen finds the "right" cell, it stimulates it to enlarge, divide, and produce a family, or clone, of antibody-forming cells.*

the tiny amount of antibody made by a single cell. Together with Dr. Joshua Lederberg and later Dr. Olavi Mäkelä, I devised methods by which single cells from antibody-forming organs, such as the lymph node or spleen, could be micromanipulated into tiny, microscopic droplets of nutrient fluid and incubated at body temperature for several hours. With such an experimental design, the total amount of antibody manufactured might be minute, but the concentration achieved could be reasonably high.

We tested the antibody content by introducing fast-swimming bacteria into the microdroplets, each of which contained a single cell (Figure 3). If antibody was present, the bacteria stopped swimming within seconds, and eventually became aggregated into clumps. Without antibody, they went on swimming indefinitely. With this method, we showed that when an animal was immunized with bacterial strains A, B, and C, the cell population as a whole made antibodies against all three strains, but each individual cell made *only* anti-A *or* anti-B *or* anti-C. We coined the rule of "one cell-one antibody," which since has been amply confirmed. Also, much simpler methods of studying antibody secretion from single cells have been developed since.

The one cell-one antibody rule was consistent with clonal selection, but certainly did not prove it. It took a decade's research not only in our laboratory but in dozens of others throughout the world to achieve both proof and considerable elaboration. One crucial experiment in this series, performed by my colleagues G.L. Ada and P. Byrt, is shown in Figure 4. It is now clear from many other lines of evidence that lymphocytes do have antibody receptors on their surfaces that allow the cells to recognize and react with antigens, and that the essential features of Burnet's theory are correct.

THE STRUCTURE OF THE ANTIBODY MOLECULE

An influence even more seminal for the development of immunology has been the growth of our knowledge on the structure of the antibody molecule. This work was

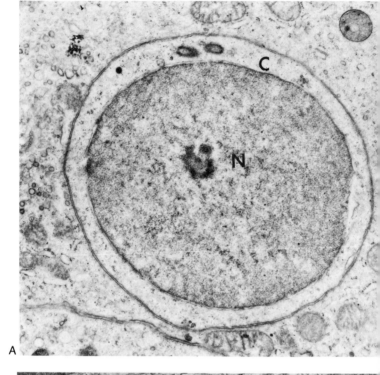

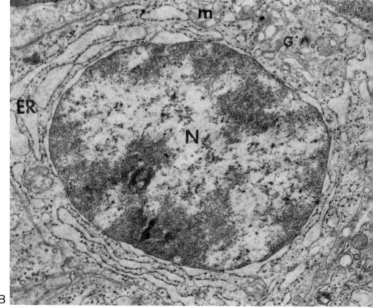

pioneered by Drs. R.R. Porter of Oxford University and G.M. Edelman of The Rockefeller University, who shared the Nobel Prize for Medicine in 1972. A great number of other protein chemists also have made distinguished contributions. Two factors slowed down the search for an exact molecular picture. First, antibody is a relatively large protein, and each molecule is made up of separate subunits, called chains. Second, as antigens stimulate more than one lymphocyte to divide, the antibody poured into the serum is polyclonal—a highly impure mixture of protein molecules with different structures. Most of the detailed structural work has, in fact, been done on the product of a cancerous, antibody-forming cell. Some humans develop a disease called multiple myeloma, in which the progeny of a single antibody-forming cell go wild and overgrow. The antibody, or immunoglobulin, which these tumors pour out in gram quantities is pure, and it is amenable to the sorts of chemical and physical analyses that have revealed the exact structure of insulin, hemoglobin, and other proteins. There are various chemical classes of antibody, but we

---

FIGURE 2 *Electron micrographs of cells important in antibody production.*

A. A small lymphocyte. This is the unstimulated version of an immunological policeman. Its nucleus or information center (N) has the genetic machinery for antibody production, but its cytoplasm (the "factory" of the cell), labeled C, looks rather inactive. This cell does not secrete antibody, but prowls around the body waiting for an antigen to come and stimulate it. Magnification X 18,000.

B. A plasma cell. This is the end product of antigenic stimulation of cell A. One lymphocyte can divide sequentially and form hundreds of identical plasma cells. Note that the cytoplasm is much bulkier and complicated looking. Prominent features are an assembly line (ER for endoplasmic reticulum) where antibody molecules are synthesized and transported; a packaging center (G for Golgi region); and numerous powerhouses for energy production (m for mitochondria). Note that the nucleus (N) looks much the same as in A. Magnification X 15,000.

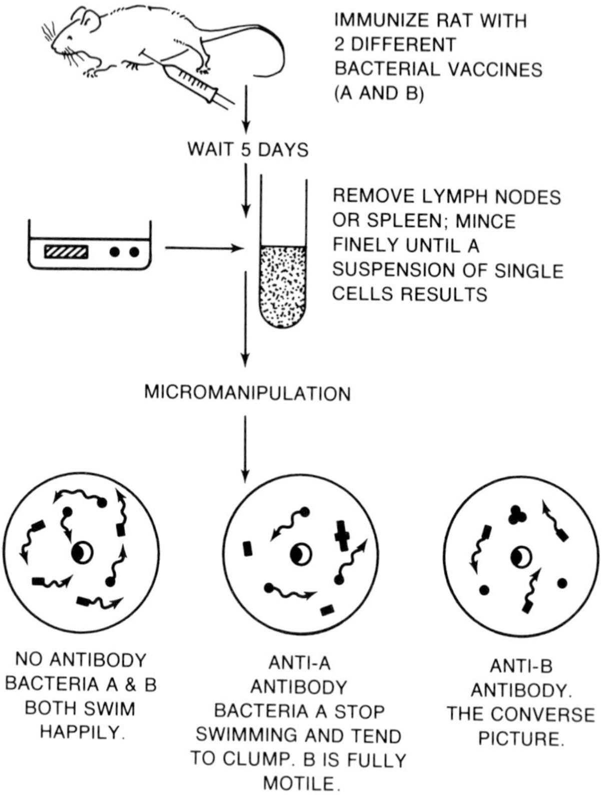

FIGURE 3 *First experimental test of the clonal selection theory. It was found that one cell always produced only one antibody.*

shall deal first with the most common variety—immunoglobulin G, or IgG, for short (Figure 5).

IgG is a T-shaped molecule made up of four subunits (called chains) bundled together. There are two identical light chains, each about 220 amino acids long, and two identical heavy chains, each about 450 amino acids long. The portion of the molecule that actually "recognizes" and unites with antigen is called the combining site of the antibody. Each IgG molecule has two identical combining sites situated near the tips of the arms of the T. Both the heavy and the light chains contribute to the combin-

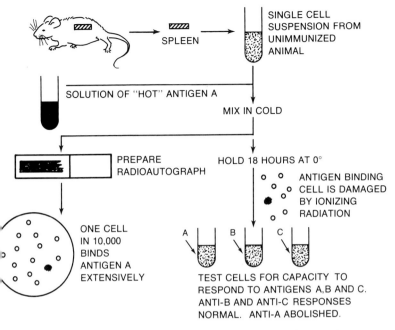

FIGURE 4 *Summary of the "hot" antigen suicide experiment of Dr. G.L. Ada and Pauline Byrt; a much more rigorous proof of clonal selection.*

ing site. This immediately yields one bit of information that helps the geneticist to understand the system. If, for example, there were 10^4 different heavy-chain genes and 10^4 different light-chain genes in an animal, each of the resultant 10^4 heavy chains could combine with each of the 10^4 light chains, yielding $10^4 \times 10^4$ or 10^8 different kinds of antibody (about 100 million). This would give an antigen a lot of diversity from which to choose!

The most remarkable feature of antibody structure becomes apparent when we analyze the sequence of amino acids in antibody chains. One-half of the light chain and three-quarters of the heavy chain follow the same rules as other proteins. In these "constant" regions, called c regions, all antibodies of a particular kind or class bear an identical amino-acid sequence. These parts of the IgG molecule are not responsible for the specific union with antigen, but are necessary to give

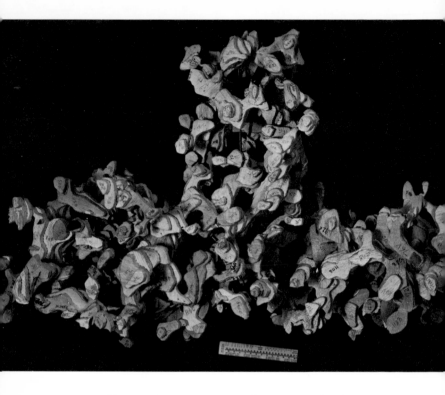

FIGURE 5 *A. A picture of an IgG molecule obtained by a powerful technique known as X-ray crystallography. The molecule resembles a 3-dimensional T. Combining sites for antigen are probably somewhere near the ends of the horizontal arms of the T, and the vertical portion represents a "handle" that does not differ from antibody to antibody.*

B. A schematic version of the subunit composition of IgG. Fab stands for the portion or fragment of the molecule which binds antigen, Fc for the portion that acts as a handle. S-S represents areas of the molecule where molecular bridges exist which either bind separate chains together, or else form loops within a given chain. CHO indicates sites of attachment of a small carbohydrate, the function of which is still unclear.

C. Schematic view of the amino-acid sequence of Ig chains derived from two different Ig molecules. Each chain has regions (either one or three) that are, like other proteins, alike from individual to individual and from molecule to molecule (c regions) and regions of equal length (v regions) that have both variant and invariant amino acids at different positions. Although only seven letters of a 26-letter alphabet are shown, each region has about 110 letters of a 20-letter alphabet.

it the most effective sort of shape, size, and characteristics. However, regions of about 110 amino acids in each of the light and heavy chains behave differently. Here, when two antibodies (e.g., anti-influenza and anti-typhoid) are compared, it can be seen that half of the amino acids will differ, as schematized in Figure 5C. These so-called variable, or v, regions of the light and heavy chains together make up the "recognizing," or antigen-binding site of the IgG molecule. It is the large number of possible permutations and combinations achievable through amino-acid substitions in these v regions that creates the huge diversity of antibody types.

Man possesses five distinct chemical kinds of an-

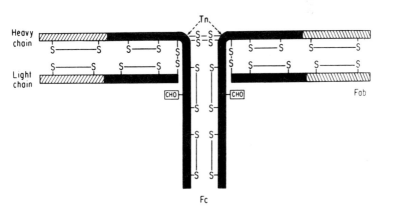

tibody. In decreasing order of their concentration in the blood stream, these are IgG, IgM, IgA, IgD, and IgE. All of these classes possess the same type of light chains and the same heavy-chain v regions. What differentiates them is the c region of the heavy chains. This confers on each class distinctive biological properties. Thus, IgG is especially good at neutralizing toxins, and it crosses the placenta readily and protects newborn babies from those infections to which the mother is immune. IgM has particular capacities leading to efficient killing of bacteria. IgA passes preferentially into secretions such as saliva, tears, milk, or the mucus lining of the respiratory or intestinal tracts. It thus acts as a kind of immunological antiseptic paint, lining various surfaces of the body. IgE activates the inflammatory mechanism, and excess production can lead to such allergies as asthma or hay fever. Despite their biological properties, the combining site of all these classes is similar, being conjointly formed by heavy- and light-chain v regions.

T AND B LYMPHOCYTES AND THEIR DIFFERING ROLES IN IMMUNITY

So far, we have confined our attention to the production of antibody molecules that are secreted by lymphoid cells, circulate through the blood stream, and exert their protective effect at some distance from the point at which they were manufactured. In effect, antibody-forming cells (Figure 2b) can be thought of as tiny secretory glands. The description "humoral immunity," or immunity caused by body secretions, has been coined to describe this type of immune response.

A second type of immune reaction is called cell-mediated immunity. Here the immune lymphocyte reacts with antigen in a strictly local setting, and the beneficial effects to the body are the result of a direct physical encounter between cell and antigen. For example, in the disease tuberculosis, the causal bacteria live chiefly *inside* the cells of the host, particularly inside scavenger cells. These cells, also termed macrophages, are important in bodily defense because they can "eat" particulate matter, whether it comes from invading bacteria or even from a splinter that is introduced into a wound accidentally. Antibody molecules have difficulty in reaching the cell-protected parasite, and so humoral immunity is ineffective in conferring protection. However, if one sets up a state of cell-mediated immunity to the tubercle bacillus, for example by the use of BCG vaccine, substantial partial protection will ensue. This is because immune lymphocytes can migrate to where any TB-infected cell is living and initiate the process of chronic inflammation, which eventually walls off the source of the antigen.

While cell-mediated immunity is clearly important in many bacterial, viral, and parasitic infections, it has been studied most intensively over the past few years for a different reason. In fact, cell-mediated immunity is centrally involved in the rejection of organ transplants and of certain cancerous tumors. Each person is genetically, and therefore chemically, different from everyone else. This means that if cells from one individual are introduced into the body of another individual, the recipient's body reacts as if it were fighting a disease-producing agent, and both humoral and cellular immune responses join the attack. Moreover, most cancers are, in a chemical sense, "foreign" to the hosts that harbor them, and they also provoke an immune attack, as we shall see more fully below. In both cases, cell-mediated immunity is the more powerful rejecting factor.

We now know that cellular and humoral responses depend on two fundamentally different types of lymphocytes, which we term T and B for short. T lymphocytes are born in the thymus, a large lymphoid organ in the front of the chest and above the heart, that is particularly large and active in childhood. The T cells are seeded out from the thymus and move to lymph nodes and spleen. When they encounter antigen, they divide and produce a large crop of activated cells capable of reacting with that particular antigen. If the antigen is on a living cell, say a kidney graft, the T lymphocyte, if unchecked, can wreak havoc, and is frequently termed a "killer" cell.

B lymphocytes come from the bone marrow, and when appropriately activated by antigen, divide and make antibody-forming cells. Thus B lymphocytes are responsible for humoral immunity. Although T and B cells look identical under the microscope, they differ in a number of subtle ways. In particular, some antigens cannot stimulate B cells by themselves. The response must be helped by activated T cells, and the mechanism of this action, in which the combined effect is greater than the sum of the individual effects, is one of the most vigorously pursued areas of immunology research today. That is because the control of T- and B-cell interaction may lie at the root of both the cancer and the transplantation problems.

ORGAN TRANSPLANTATION AND IMMUNOLOGIC TOLERANCE

If a T-cell killer attack is such a powerful way of destroying grafted cells, why are kidney transplants successful in more than 75 per cent of cases? The most important initial factor is the use of drugs that suppress both the immune reaction and dangerous inflammation. During the early postsurgery days, high dosages must be used, and while these usually keep the killer lymphocytes at bay, they unfortunately leave the patient immunologically defenseless against microorganisms, so infection is a big danger.

After some months, drug dosage can be decreased progressively, and the survival of the transplant then depends on a combination of factors. First, a mild degree of immunosuppression by drugs is maintained. Second, some cells of the grafted kidney are replaced by the patient's own cells. The third and fourth factors are known respectively as tolerance and enhancement. Immunological tolerance is a specific switching-off of lymphocytes by antigen. Its mechanism is still poorly understood, but it involves a temporary or permanent incapacitation of the reactive lymphocyte. In practice, complete tolerance is best induced by exposing very young animals to antigen and by injecting adult animals with highly soluble antigens in large doses. We cannot yet induce tolerance (or immune paralysis, as it is sometimes called) at will in human transplant recipients. However, a partial tolerance may be enough in long-retained kidney grafts. Enhancement (Figure 6) is a curious phenomenon by which antibodies directed against a graft actually help, rather than hinder, graft survival. The enhancing antibody may combine with antigen, and antigen-antibody

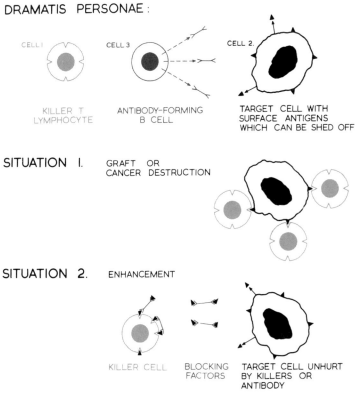

DRAMATIS PERSONAE:

CELL 1

KILLER T
LYMPHOCYTE

CELL 3

ANTIBODY-FORMING
B CELL

CELL 2.

TARGET CELL WITH
SURFACE ANTIGENS
WHICH CAN BE SHED OFF

SITUATION 1.   GRAFT OR
CANCER DESTRUCTION

SITUATION 2.   ENHANCEMENT

KILLER CELL   BLOCKING
FACTORS

TARGET CELL UNHURT
BY KILLERS OR
ANTIBODY

FIGURE 6   *The principle of immunological enhancement, with T and B lymphocytes exerting antagonistic effects on a transplanted or cancerous cell. Situation 2 depicts only one of several possible mechanisms of the action of blocking factors. Much research remains to be done in this area.*

complexes are known to be capable of paralyzing lymphocytes. Whatever the mechanism, it is clear that B cells making enhancing antibody act antagonistically to T cells acting as killers. It is probable that enhancement is a highly significant factor in allowing grafts to survive.

We must not neglect the importance of genetic factors in transplant survival. It is possible to type the chief antigens on cells that induce a strong immune response. They constitute a complex set of molecules, but if good matching is achieved between donor and host, the chances of graft survival are improved greatly. In practice, such matching is most frequently obtained by using a brother or sister of the patient as the kidney donor.

Transplantation of other vital organs, such as heart, liver, or lung, using cadaver donors, is performed, but less frequently and with a lower success rate than in kidney grafts. Various factors account for this. It is usually more difficult to get such patients into good pre-operative condition, because no artificial hearts or livers, comparable to the artificial kidney, have been developed. Therefore, the operations are more demanding technically. The organs, and particularly the blood vessels supplying them, may be more susceptible than the kidney to immunological damage. Also, transient failure of function, permissible in a transplanted kidney, would be fatal in a heart or lung. Nevertheless, the mere fact that several two-year survivors of heart transplants are still alive indicates that the procedure is feasible. Now we must find ways to overcome the immunological barrier more fully.

AUTOIMMUNITY

More than 70 years ago, the great German biologist Paul Ehrlich speculated on the chaos that would result within the body if a major immune attack were mounted against some internal, or "self," component. In fact, the immunological tolerance that we induce artificially to test antigens is only a model for a process that goes on within us all the time. Our own lymphocytes are constantly being made tolerant to proteins and other antigens present in our blood. Therefore, although our own red blood cells, for example, may cause an immune response when injected into another person, we ourselves do not normally make antibodies against them. Sometimes this tolerance of self components breaks down, and people do develop antibodies and cell-mediated immunity to one or more of their own organs, such as thyroid, stomach lining, or liver. This is the source of the term autoimmunity, which means an immune reaction against self. The result, of course, is disease, which may be serious or even fatal. Such autoimmune diseases include many forms of arthritis, blood disorders, chronic kidney and liver ailments, and a majority of the illnesses that affect the endocrine glands. Broadly speaking, autoimmunity represents a failure in immune regulation. Scientists are still puzzling out the nature of the tolerance breakdown; one factor of importance is an inherited predisposition. The treatment of autoimmune disease follows much the same principles as that for organ-graft rejection.

CANCER IMMUNOLOGY

The overgrowth of cells, which we term cancer, is the end result of complex processes that we are barely beginning to understand. It is unwise to discuss "the cause" of cancer, as we already know that a vast variety of different triggering factors are involved. Therefore, it is unrealistic to expect some sudden breakthrough that will solve all aspects of the cancer problem once and for all. Nevertheless, there is great excitement in cancer research at the present time because of certain advances in tumor immunology. These relate to prospects for both early diagnosis and better treatment.

The biochemical abnormalities of cancer cells are numerous, but in many cases, molecules are produced to which the body is not immunologically tolerant. These appear at the surface of the cell as "tumor-specific antigens," or sometimes they are secreted in sufficiently large amounts to reach the blood stream, in which case sensitive techniques can detect them in the serum. Sometimes these antigens are the end result of the action of a cancer-producing virus. In other cases, they reflect a reversion by the cancer cell to a growth and behavioral pattern characteristic of embryonic tissues, which grow much more rapidly than they do after birth. In many instances, the nature of the antigen is unknown. The first important point is that such antigens may be used to aid early diagnosis of cancer. This is particularly true for those antigens which enter the blood. A simple blood test, where a routine yearly check-up could give an early warning, would be a significant step forward in the fight against cancer. This is not yet a reality, but may become possible in time.

The second great hope is that it may be possible to strengthen the body's immune attack against cancer cells. Probably many precancerous cell clusters are rejected by the lymphocytic policemen without anyone being aware of it. In fact, this immune surveillance function may be one of the prime duties of T lymphocytes. If this is indeed the case, only those tumors that slip through the net ever become clinical problems. Clearly, any immune attack against cancer is a numbers game —so many killer lymphocytes versus so many cancer cells. If this balance could be tilted in the lymphocytes' favor by some form of tumor-antigen vaccine, a new weapon would be in our hands. It is very hard to get rid of every last cancer cell, as we know from the results of drug or radiation treatment, and so immunotherapy might find a place after the bulk of the tumor mass is removed by one of the present methods. Also, there are nonspecific ways of stimulating lymphocytes, and these also may be brought into play.

The biggest bugbear to a vigorous exploration of these possibilities is the strangely paradoxical role of B lymphocytes in cancer progression. Most cancer patients, at least at an early stage of their disease, do possess some killer lymphocytes against their own cancer, but blocking factors, probably complexes between enhancing antibody (produced by B cells) and tumor antigens, co-exist in the blood and antagonize the action of the killer T cells. If we were to find a specific stimulator of T-cell activation, we might find that T-cell killer function is increased (as desired), but that the T helper function is also increased. The end result might be the increased production of "bad" antibodies by B cells and a stalemate in the war against cancer. Therefore, it is clearly vital to obtain much deeper insights into the basic biology of the T and B lymphocyte systems.

OTHER AREAS OF CLINICAL IMMUNOLOGY

Immunology is infiltrating many other areas of clinical medicine. Allergy and hypersensitivity are caused by immune processes, as are many complications of viral and bacterial infections—for example, rheumatic fever, acute nephritis, or the kidney complications of malaria. Too much antibody formation can be a bad thing, and the regulation of the immune system is so delicately balanced that not infrequently it goes wrong. A major new medical specialty, termed immunopathology, has grown up in the last decade to deal with all these clinical implications of fundamental immunology.

We have not said much about the newer conventional vaccines, such as those recently developed against mumps and measles. These represent technical triumphs, without involving any new principles. However, there are still many challenges in this area of immunology. For example, we have no vaccines against infectious hepatitis, venereal diseases, or the serious tropical maladies such as malaria, leprosy, or parasitic diseases. Indeed, it might be proper to place rather more emphasis on research into these areas, less "glamorous" than transplantation or cancer research, but enormously important for the health of mankind.

CONCLUSIONS

The new immunology obtains much of its excitement from its interdisciplinary character. Few branches of biomedical research impinge on so many important subjects. Those of us working on the basic biology of the lymphocyte see a goal much greater than the solution of an interesting puzzle. We hope our progress will fertilize a whole range of fields, both fundamental and clinical. Most scientists spend the bulk of their time attacking some highly specialized problem, but deep in their consciousness is the hope that an elegant and simple solution of a particular problem will exert a "domino" effect, a cascade of problem-solving across a wide front. The great challenge in science is to achieve a balance between depth and breadth.

The goal of a linear continuity of knowledge between the most fundamental level of science, say mathematics or physics, and the most applied, such as social medicine or political science, is clearly beyond any single human brain, yet civilization as a whole must strive to reach this goal. In what is, as yet, a preliminary and struggling fashion, modern immunology has fulfilled a bridge-building function. It has been an exhilarating beginning, but, like most of my colleagues, I find myself humbled by the contrast between past achievements and the immensity of residual ignorance. The hope for an extension of the fine edifice lies in the large army of gifted younger workers who have come into the field in recent years. They are overawed neither by the triumphs of our established leaders nor the daunting nature of the remaining problems. Perhaps some reader of this short summary will wish to join the fray, for great adventures lie ahead.

CHAPTER 9 DAVID S. HOGNESS

Human Use of Genetics

Living organisms are the greatly magnified expressions of the molecules that compose them.
George Wald, in *Evolutionary Biochemistry*

SOME TIME AGO I became interested in retrievers and sought out an old man who had spent much of his life training these remarkable dogs. Near the end of our conversation he asked about my own work, and upon finding that it was somehow connected with genetics, burst out that this was the most interesting thing I had said that evening. Like most of us, he was familiar with the word genetics, and used it in two ways. On the one hand, it was a convenient term to cover the commonsense knowledge that men have had for ages concerning hereditary stability and variation: that while dogs yield dogs and never cats, different dogs exhibit different characteristics which sometimes can be combined in their progeny. By contrast, he also had an almost mystic idea that some people (geneticists) had a fund of knowledge (genetics) above the common-sense level, and that if *this* genetics were properly applied to the breeding of dogs, unusual animals could be obtained, which in his dreams would bring home the bacon with little or no training.

His two usages correspond in a rough way to two periods in the development of our knowledge of heredity: a very long pregenetic period, including all the years before 1865—that is, before Mendel's definition of the gene—and the half-century after the rediscovery in 1900 of Mendel's experiments, the period of classical genetics. About the extraordinary recent burst in knowledge that we call molecular genetics, he hadn't an inkling.

Although somewhat arbitrary, these divisions have a certain use in considering the effects of this knowledge on man and his environment. Most obvious are the effects of the pregenetic period. Indeed, it was the early recognition that like begets like and of the possibility of selective breeding that resulted in man's developing domestic animals and crop plants and allowed him to make the transition to the agricultural and urban societies that ushered in the first civilizations. Without an accurate understanding, but with a strong sense of direction, man began to tap the pool of genes in his environment to his own ends. It was a slow and mostly haphazard process, but it worked. It worked because man unwittingly applied that very process upon which the evolution of all living forms depends—the application of a selective pressure, or screen, on an existing set of genes, which, by eliminating some members, changed the nature of the set. The enormous difference was that man had begun the design of these screens.

NEW MODELS FOR OLD

It was an extremely slow process, because the available models for the mechanics of heredity were inaccurate; nor did they improve with time during the pregenetic period. Thus the model adopted by Darwin in the nineteenth century was little different from the model taught by Hippocrates in the fifth century B. C. The trouble was that a selective screen of experimental tests was never applied to the various models of this period. As soon as Mendel and the first geneticists of the classical period devised relatively simple experimental test systems, a revolution in hereditary concepts occurred; pregenetic models were discarded and replaced by a formal genetic model of considerable accuracy.

The model of classical genetics is based, first, upon the concept of separate and distinct units of hereditary information called genes, each of which is reproduced unerringly from generation to generation; second, upon the chromosome, in which many different genes are assembled in a linear array, each gene occupying a particular position; and third, upon mutation, a rare, random process whereby the information content of a gene (or its position) is altered, and the altered form is then reproduced in the same unerring manner as its progenitor. Hence the hereditary information available to a given individual resides in the total set of genes that it contains. This set is divided into subsets according to the numbers and types of chromosomes obtained from each parent. The laws for the transfer of genetic potential from one generation to the next will therefore depend upon how the chromosomes in each parent are transmitted to the offspring. We need not discuss these patterns of transmission in detail to appreciate that, because the patterns are simple and general, genetic laws are simple and general.

This generality was unique in biology, for the genetic laws discovered by the classical geneticists provided the first coherent, unifying principle for life on earth, or at least for the higher forms of that life. (Bringing the lower forms, such as viruses and bacteria, into a common framework with the higher forms had to await the development of molecular genetics.) Indeed, one of the most striking examples of the societal values of basic research is that the genetic rules derived from studies on such experimentally convenient sources as the fruit fly are applicable to man and to the domestic animals and plants upon which he depends.

APPLICATIONS OF THE GENETIC MODEL

The practical consequences of an accurate model for heredity have been so numerous and varied that we often fail to appreciate that they derive from this basic research. In agriculture, the model has allowed the logical construction of more sophisticated and efficient selective screens to create new gene combinations, thereby increasing the specificity and the rate of acquisition of new varieties. For instance, the yield from different crops has been augmented by selecting for resistance to particular pests or for growth cycles to fit different growing seasons or for advantageous shapes and structures, sometimes with dramatic effects. Short, stiff-strawed varieties of rice developed at the Rice Research Institute in the Philippines increased the yield from 1,200 pounds to more than 9,000 pounds per acre. This extraordinary increase soon allowed the Philippines to become an exporter of rice, and we may expect that the food supply will be profoundly altered as these strains are introduced throughout southern Asia.

The specific characteristics that have been selected in response to man's varied requirements are astonishing. Seedless watermelons are now available to those with inhibitions about spitting, and pineapples have been developed that will fit a number two can with a minimum of waste. These examples, among hundreds, indicate why we believe that, even now, man has only begun to tap the reservoir of genes made available by centuries of mutation and natural selection.

And what of man? One consequence of the application of classical genetic knowledge to humans has been the belated recognition that a considerable number of human ailments have genetic origins. About 2,000 such disorders are now recognized, and that number will undoubtedly increase as we obtain better techniques of recognition. And as we become more capable of preventing or curing the diseases caused by such infective agents as bacteria and viruses, genetic disorders will inevitably occupy an increasing fraction of our attention. Even now it is estimated that 25 per cent of our hospital beds are occupied by those suffering from genetic disorders.

There are two ways of coping with genetic disorders—prevent their occurrence or devise some treatment. The harsh facts are that the only way to prevent the occurrence of individuals with genetic disorders is to prevent their conception or their birth. If they are born, there is at present no complete cure in the sense of replacing the defective gene with its normal counterpart (although later I shall consider a possible method for achieving something close to this.) The better alternative is generally prevention.

The new profession of genetic counseling derives from the preventive possibilities. Prevention prior to conception requires that defective genes in the parents be detected, and that may be difficult. However, a good example of how carriers of a genetic disease can be detected is provided by sickle-cell anemia. This disease is caused by a defect in one of the genes that determine the structure of hemoglobin, a protein in the red blood cells that plays a vital role in the transport of oxygen. The defective gene causes a minute change in the amino-acid sequence (which I discuss later in more detail) of the protein, and this structural alteration in turn leads to a complex set of changes that fatally decrease the ability of the red blood cells to transport oxygen to the tissues. The disease is called sickle-cell anemia because the normally round red blood cells become sickle-shaped and fragile as a result of these changes.

In certain populations, particularly in Central Africa, the gene for sickle-cell anemia is present in 20 per cent or more of the population. The individuals who carry the sickle-cell gene in these populations are, for the most part, perfectly healthy, because each defective gene is paired with a normal gene. These individuals are said to have sickle-cell *trait*. However, if such a person marries another individual with sickle-cell trait, the chances are one in four that a child born to them will have the fatal sickle-cell disease itself (Figure 1).

Now there are ways to identify persons who carry the sickle-cell trait. For example, their red blood cells, which have the normal round shape in the body, will adopt the sickle shape under certain laboratory conditions, whereas the cells from normal individuals who do not carry the sickle-cell gene will not. Alternatively, the hemoglobin extracted from the red cells can be examined for the presence of the altered hemoglobin molecules that are formed from the sickle-cell gene, as illustrated in Figure 1. Theoretically, at least, all suspected persons can now be tested for the trait and so be better informed on planning their families.

If we take into account the fact that the development and maintenance of the human body depends upon the

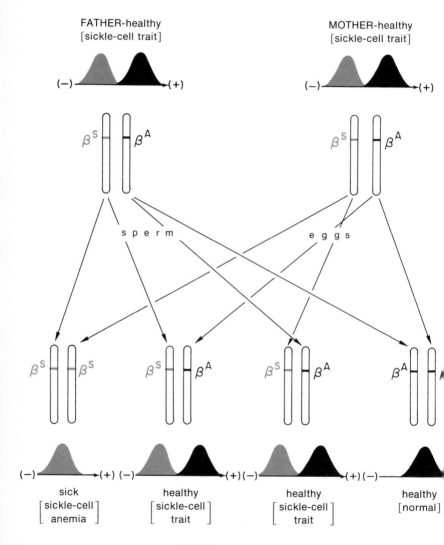

KINDS OF CHILDREN

precise function of thousands of different proteins whose structure is determined by thousands of different genes, we can understand why it is thought that most genetic disorders result from a change in either the structure or the rate of synthesis of specific proteins. And indeed we now know of many examples in addition to sickle-cell anemia that confirm this conclusion. Hence, the operative rule for the detection of a defective gene has become "Cherchez la protein!" If the altered protein that is responsible for a genetic disorder can be identified, it is often possible to design simple, quick tests for its presence, as in the case of sickle-cell anemia and a few other genetic disorders. This, in turn, allows the detection of the defective gene, even in parents who are healthy but carry a faulty gene.

Having detected the presence of defective genes in the parents, the genetic counselor might then say to them something along the following lines. "Here's how we

---

FIGURE 1   *Diagram illustrating the kind of children expected when both parents have sickle-cell trait. The paired verticle "rods" indicate the combination of β^S and β^A genes in the relevant chromosome pair. Each sperm from the father (or egg from the mother) contains only one chromosome from the pair, and since this chromosome is randomly selected, just as many sperm (or eggs) contain the mutant gene (β^S) as contain the normal gene (β^A). Random pairing of these chromosomes by mating of the parents yields three kinds of children [β^S/β^S (sickle-cell anemia), β^S/β^A (sickle-cell trait), and β^A/β^A (normal)] in the ratio 1 : 2 : 1. Hence, there is one chance in four that a child will exhibit the anemia; two chances in four that, although healthy, he will exhibit the trait; and one chance in four that he will be normal in respect to this genetic disorder.*

The types of hemoglobin formed by the parents and children are indicated by the bell-shaped curves. The hemoglobin formed by the β^S, or sickle-cell gene, and that formed by its normal counterpart, β^A, move from left to right at different rates in the electric field indicated by the (−) and (+) symbols. They therefore occupy the different positions indicated by the bell-shaped curves after a given time of exposure to the field.

stand. The *chance* that your child will be a genetic cripple is one in four. If he is, we can reduce the effects of the disorder by treatment, but we can never effect a complete cure. Even with the best treatment, there is a considerable *chance* that the child's abilities will be impaired." Although the scenario may be more or less drastic than this, the general rule is the same—specific factors of chance play a dominant role in the choice given the parents prior to conception.

Clearly, these chance factors would be reduced if the parents could be given a choice *after* conception. This is now possible in a small but growing number of cases as a result of new procedures that allow fetal cells obtained from the amniotic fluid to be analyzed genetically. If a defect is found, the parents then have the choice of abortion. For instance, a visual examination of the chromosomes in such cells may reveal a condition in which one of the 23 types is represented three times rather than two—a condition that is known to lead to severe mental retardation and a variety of physical abnormalities called "mongolism," or Down's syndrome. Mongolism occurs in about one birth in 700, and about one-third of the patients in institutions for the most severely retarded suffer from this disorder. Humanitarian concerns for the child, for its family, and indeed for society, all act in concert to favor abortion in this case.

But in other cases, where the genetic disorder is less severe, the decision is not so simple. For example, there is a significant but not invariable correlation between males carrying an aberrant combination of the sex chromosomes and marked antisocial behavior. This combination is designated XYY and indicates that these rare males contain one X chromosome and *two* Y chromosomes. Since normal males contain one X paired with *one* Y, and normal females contain a pair of X's but no Y, this aberrancy represents another case in which the chromosomes in one of the 23 pairs, in this case the sex pair, are overrepresented.

What will be the response of parents when told that the

fetus contains the abnormal XYY combination? What should be the response of society? Clearly, the personal desires of the parents will often be at odds with the perceived interests of society. When discrepancies of this nature become obvious, societies generally react in their own favor, and to my mind the relevant question is not whether societies *will* act, but when and in what manner.

Because we shall be exposed to both personal and societal decisions of this specific nature with increasing frequency, there is, in my opinion, some urgency in preparing a proper base by moving genetics to the center of our educational experience. And not only for the reasons discussed thus far. The knowledge we have acquired recently regarding the molecular nature of genetic information forms the unifying link and fundamental basis for all terrestrial life, and indicates that man's potential for changing that life through the use of his intellect is far greater than we would have guessed even thirty years ago.

THE UNIFYING PRINCIPLES OF MOLECULAR GENETICS

What has happened in that short period? It is much as if we had taken a drink from Alice's elixir, and had shrunk smaller and smaller, until we could see the individual atoms of which the genes are composed. Such a vision is startling, for we see immediately that the array of atoms is extraordinarily simple. Two strands of linked atoms are twisted around each other to form a long, thin rope, and clusters of atoms are attached to each strand at short, regular intervals (Figure 2A). Were we to walk along this ropelike molecule and examine each of the thousands of these atomic clusters carefully, we should come to the remarkable conclusion that there are only four kinds of clusters, or *bases*, as we shall refer to them. Moreover, it would be apparent that the two twisted strands in the genetic molecule, or DNA (deoxyribonucleic acid), are held in exact register, one relative to the other, by very specific interactions between the

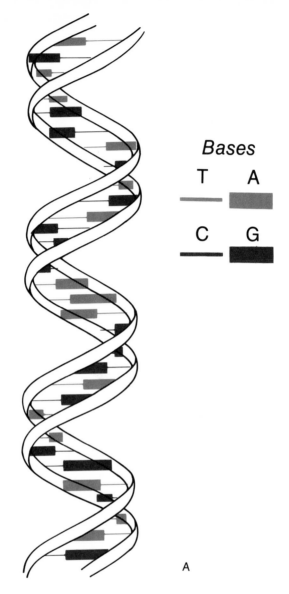

Bases

T   A

C   G

A

FIGURE 2 *Diagrams illustrating (a) the structure of the genetic material, or DNA, and (b) the use of the Watson-Crick pairing rules to form two "daughter" molecules that have the*

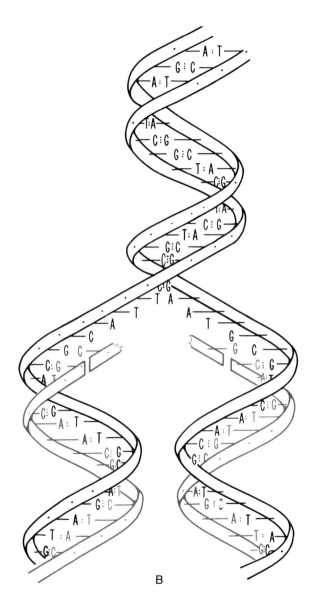

B

same structure as the "parent" molecule during the replica-
tion of DNA. The four bases are designated by T (thymine), A
(adenine), C (cytosine), and G (guanine).

bases in each—a given base on one strand interacts, or pairs, with only one type of base on the other. This interaction can be described by two simple pairing rules. Designating the four bases by A (adenine), G (guanine), T (thymine), and C (cytosine), we see (Figure 2B) that A always pairs with T (and vice versa) and that G always pairs with C (and vice versa). These two simple rules were deduced by Watson and Crick some twenty years ago, and, as we shall see, they determine most of the properties of genes.

THE FIRST UNIFYING PRINCIPLE. One of these properties is the ability of genes to be reproduced or replicated. In classical-genetic terms, this replication implies that the informational content of the gene be duplicated without error. In molecular-genetic terms, it means that the sequence of bases along the strands of the DNA be reproduced exactly in the new molecules, because, as will become apparent, the genetic information resides in these sequences.

The accurate replication of genes depends upon a strict adherence to the simple pairing rules. Let me illustrate this by asking you to imagine that the two strands of the DNA have become separated along a short part of their length, as is shown in Figure 2B. Now suppose that we begin to construct two new strands by using each of the separated, original strands as templates, or patterns, to which we apply the pairing rules. Thus, we place a G in a new strand opposite a C in the old, a C opposite a G, an A opposite a T, and a T opposite an A. If we continue this process, untwisting the original pair of strands as we proceed, the result will be the construction of two "daughter" molecules, each containing base sequences identical to the original, or "parent," molecule. This is so because each of the daughter molecules contains one strand from the parent molecule, and the pairing rules tell us that the base sequence in one strand exactly determines the sequence in the other.

This accurate process for duplicating DNA not only explains how the sequence of bases in a given gene is

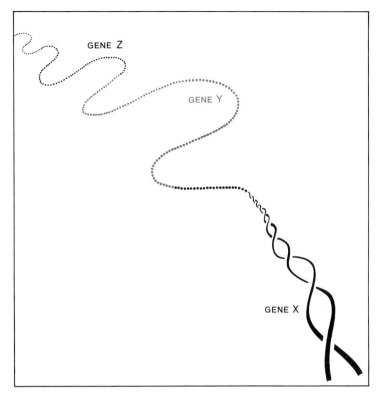

FIGURE 3 *Diagram illustrating how many different genes are arranged in a single DNA molecule to produce a linear sequence of genes within a single chromosome.*

reproduced, but also explains how the linear array of genes in a chromosome is maintained from generation to generation. The gene order within a given chromosome is maintained because its DNA consists of one exceedingly long DNA molecule that contains all of the genes in linear order (Figure 3). Each gene occupies a segment of the DNA molecule, one gene, or segment, following another along the length of the DNA. The replication of this giant chromosomal DNA must then yield two daughter molecules that contain exactly the same gene sequence as that found in the parent—just as each gene in the daughter molecules must contain exactly the same base sequence found in the parent.

Now the important point is not only that the replication of the linear array of genes in a chromosome can be explained by simple molecular rules, but also that the same rules apply to the replication of *all* genes, whether in man, peas, fruit flies, or bacteria. We shall call this the first unifying principle of molecular genetics. It means that if we could somehow insert a segment of DNA that contains a gene from one life form, say bacteria, into the DNA in a chromosome of another, say human, we should expect the bacterial gene to be replicated along with the human genes in that chromosome.

THE SECOND UNIFYING PRINCIPLE. How can the linear sequence of bases in a segment of DNA provide the genetic information specific to a single gene? The kind of genetic information we know most about is that provided by the genes which dictate the linear sequence of amino acids in proteins. There are thousands of different genes of this kind in each cell-producing organism, from bacteria to man; even in viruses, such genes occupy a majority of the DNA. If we restrict our consideration of genetic information to this ubiquitous class of genes, the above question takes a more specific and useful form: How can the linear sequence of bases in a segment of DNA determine the linear sequence of amino acids in a protein chain? It is the answer to this question that provides us with the second unifying principle of molecular genetics.

It is possible to appreciate this second principle without a detailed knowledge of the complex molecular mechanisms that are involved in the translation of one linear sequence into another. Indeed, there is a certain advantage in stripping the answer to the above question of its mechanistic details in order to emphasize the simplicity of the formal rules relating the two sequences. These rules prescribe an amino-acid sequence for any given base sequence, so they are known collectively as the genetic code. All of the available evidence indicates that there is only one genetic code, which is used by all the different life forms on earth. This universality constitutes the second unifying principle.

The basic steps in the decoding of a genetic base sequence are outlined in Figure 4. Because a given segment of DNA contains two related but different base sequences (one per strand), the first step involves the selection of the correct one. This is accomplished by making a copy, or transcript, of the sequence in only one of the two strands, using the other strand as a template to which the Watson-Crick pairing rules are applied (Figure 4A). Although this transcription bears an obvious resemblance to DNA replication, it is the differences which are to be emphasized here. First, the single-stranded product has a slightly different chemical composition from DNA, which we note by calling it RNA (for ribonucleic acid). Whereas three of the four bases are the same in both nucleic acids (C,A,G,), the base U (for uracil) in RNA substitutes for T in DNA. This substitution results from the fact that U is so similar to T in structure that it obeys the same pairing rule—that is, during formation of the RNA transcript, a U is paired opposite an A in the DNA template (Figure 4A).

Transcription also differs from DNA replication in that only a segment of the chromosomal DNA is copied. This allows different genes (or different clusters of adjacent genes) from the same chromosome to be transcribed at different times and in different cells within the same organism. Moreover, the single-stranded RNA transcript separates from the chromosomal DNA and therefore can function as the template for the sequential assembly of amino acids into proteins at positions and times that are independent of the chromosomal state.

The translation of the base sequence in RNA can be viewed as reading a message made from a four-letter alphabet (U,C,A, and G) arranged into three-letter code words that designate the individual amino acids. Starting at a particular position, the rule is to read the adjacent, nonoverlapping code words one after the other, proceeding down the sequence in a direction determined by the properties of the RNA. In the example given in

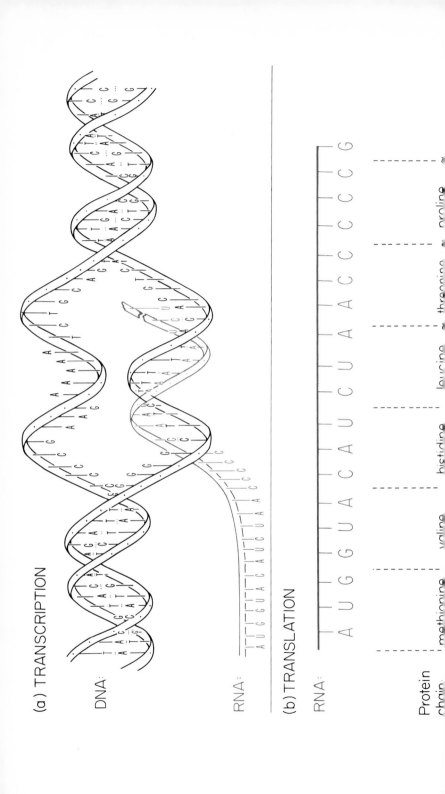

(a) TRANSCRIPTION

DNA:

RNA: AUGGUACAUCUAC

(b) TRANSLATION

RNA: AUGGUACAUCUAACCCCG

Protein
chain: methionine — valine — histidine — leucine — threonine — proline —

Figure 4B, the first code word, AUG, represents the amino acid methionine, which therefore occupies the first position in the protein chain. The second code word is GUA and represents valine, which is then linked to methionine to occupy the second position in the chain. This reading process is continued, adding one amino acid after another, until the translation is terminated by a special punctuation code word.

Because a four-letter alphabet can generate 4^3, or 64 three-letter code words, a complete dictionary for the genetic code consists of sixty-four entries, each with its definition. The construction of this dictionary—given in Table I—was essentially complete by 1966. Only three code words (UAA, UAG, UGA) do not represent any amino acid, and for this reason were designated as "nonsense" words. However, it soon became apparent that, by ending translation, they act as the punctuation signals mentioned above. Each of the remaining sixty-one code words designates one, and only one, amino acid.

The genetic code is therefore not ambiguous—the translation of the base sequence in a gene yields one, and only one, amino-acid sequence. Taken in conjunction with the second principle of a universal code, this means that, upon translation, a given gene should yield the same protein chain in any cell in any organism. Returning to our hypothetical example of a bacterial gene inserted into a human chromosome, we now conclude that, in addition to being replicated, such a gene, if translated, should yield the same protein chain in the human as in the bacterial cell.

FIGURE 4 *Diagram (left) shows how a linear sequence of base in one strand of DNA is (a) first transcribed to yield the same sequence in the messenger RNA (with U substituted for T), and (b) then translated into an amino-acid sequence in a protein chain by reading the three-letter code words in the RNA.*

TABLE I. *Dictionary for the Genetic Code*

SECOND BASE

		U	C	A	G
FIRST BASE	**U**	UUU ⎱ = Phe UUC ⎰ UUA ⎱ = Leu UUG ⎰	UCU ⎫ UCC ⎬ = Ser UCA ⎪ UCG ⎭	UAU ⎱ = Tyr UAC ⎰ UAA = term. UAG = term.	UGU ⎱ = Cys UGC ⎰ UGA = term. UGG = Try
	C	CUU ⎫ CUC ⎬ = Leu CUA ⎪ CUG ⎭	CCU ⎫ CCC ⎬ = Pro CCA ⎪ CCG ⎭	CAU ⎱ = His CAC ⎰ CAA ⎱ = Gln CAG ⎰	CGU ⎫ CGC ⎬ = Arg CGA ⎪ CGG ⎭
	A	AUU ⎱ = Ile AUC ⎰ AUA ⎰ AUG = Met	ACU ⎫ ACC ⎬ = Thr ACA ⎪ ACG ⎭	AAU ⎱ = Asn AAC ⎰ AAA ⎱ = Lys AAG ⎰	AGU ⎱ = Ser AGC ⎰ AGA ⎱ = Arg AGG ⎰
	G	GUU ⎫ GUC ⎬ = Val GUA ⎪ GUG ⎭	GCU ⎫ GCC ⎬ = Ala GCA ⎪ GCG ⎭	GAU ⎱ = Asp GAC ⎰ GAA ⎱ = Glu GAG ⎰	GGU ⎫ GGC ⎬ = Gly GGA ⎪ GGG ⎭

term. = "nonsense" code words that cause termination of translation.
Abbreviations for amino acids:

Ala:	alanine	Gly:	glycine	Pro:	proline
Arg:	arginine	His:	histidine	Ser:	serine
Asn:	asparagine	Ile:	isoleucine	Thr:	threonine
Asp:	aspartic acid	Leu:	leucine	Try:	tryptophan
Cys:	cysteine	Lys:	lysine	Tyr:	tyrosine
Gln:	glutamine	Met:	methionine	Val:	valine
Glu:	glutamic acid	Phe:	phenylalanine		

NEW GENE COMBINATIONS

The two unifying principles are recent acquisitions, and only now are we beginning to appreciate some of their radical implications. In particular, the principles offer the possibility of breaking down the species barrier in man's continuing search for new, advantageous gene combinations.

Let me illustrate this possibility by an example taken from our laboratory. It centers around certain viruses that have the capacity to insert their small set of genes, contained in a single DNA molecule, into the larger set of genes contained in the DNA of a chromosome in the host cell. Once inserted, the viral genes are replicated along with the host genes. The effect is to create a new line of cells, each of which contains the added viral genes. In the case at hand, the host cells are bacteria, but similar viruses exist in animal cells and we may reasonably suppose their existence in plant cells.

Given such natural systems for genetic integration, we ask whether the viral DNA can carry and insert into another chromosome whatever foreign DNA we choose. Suppose we attach, by a series of highly specific biochemical reactions carried out in a test tube, a small segment of foreign DNA to the viral DNA and then cause this hybrid DNA to be integrated into a chromosome of the host cell. For reasons unrelated to the point I wish to stress, we are trying by this means to insert specific small sets of genes from fruit flies into bacteria. The procedures that are being developed are quite general, so it should be possible not only to create hybrid bacteria containing genes from fruit flies or frogs or even humans, but also to create hybrid animal or plant cells containing genes from quite different species.

The implications of this type of genetic mixing are indeed radical. Although too many questions remain unanswered to predict accurately the specific applications that we may expect, certain possibilities come easily to mind. One is the treatment of genetic defects in man. Here, we imagine the insertion of new genes into the cells of genetic cripples in order to provide the functions that cannot be contributed by the defective genes.

Ideally, the added genes should be the normal human counterparts of the defective genes. Before they can be added, the DNA molecules that contain them must first be isolated. Can we so isolate specific human genes? This may indeed be possible in the immediate future, thanks to certain methods that are just now becoming available. However, there are other possibilities. It may be, as I have indicated, that equivalent genes isolated from other species will serve as well. For example, the human genetic disorder known as galactosemia results from the loss of a particular protein that catalyzes a reaction in the metabolism of a sugar, galactose. This same reaction is also catalyzed by a protein in bacteria, and DNA molecules that contain the gene for the bacterial protein can be isolated easily. We might therefore hope to insert the bacterial gene into certain cells of a galactosemic individual and thereby relieve the disorder. As yet, this is only a hope, but future studies might make it a practical possibility.

To avoid misunderstanding, one aspect of these designs for genetic therapy must be emphasized. The targets for the genetic insertion are not the germ-line cells that give rise to the egg or sperm. Rather, they are one or more of the other kinds of differentiated cells in the body—that is, the somatic cells used to construct the nervous system, the heart, or indeed all tissues except those generating the egg or sperm. Hence, successful genetic therapy of this kind would relieve the disorder *only* in the treated individual. The inserted genes, not being in the germ-line cells, would not be transmitted to the progeny, whereas the defective genes would.

The insertion of genes into germ-line cells is an extraordinarily more complex matter and the incentives are quite different. Here the aim would be to create novel variants of an existing species. I can imagine attempts to create particularly advantageous domestic plants or animals, but I do not find that schemes to create supermen by such techniques are credible. This is because foreign genes that are present in the fertilized egg will be replicated so that they are present in all cells during the development of the individual. These added genes may therefore be expressed at any time and in any tissue, or they may derange the time and place of expression of other genes. The orderly development of an individual results from an exquisite and highly integrated system for regulating the expression of different genes at different times, so the possible effects (both desired and undesired) of the foreign genes would have to be tested at all stages of development—from the beginning embryo to the adult human. Who would allow—who would want to perform such potentially horrendous experiments on man?

By contrast, there should be no hesitation in trying to create novel plants of particular advantage to man. For example, we can imagine creating plants that would not need nitrogen fertilizers. This might be done by inserting bacterial genes that allow the needed nitrogen to be obtained from the air, rather than from fertilizers in the soil. Here, the aims are better defined, the system is simpler, and ethical objections, if they exist, are rare and tenuous.

Clearly, many basic questions must be answered before we can determine if any such ventures are truly feasible. Some of those answers will undoubtedly restrict our imagination; others will open it to new domains. On balance, I am convinced that we are on the eve of a revolution in man's search for new gene combinations, a revolution that I expect will unfold in the next few decades. Except in the area of genetic therapy, the practical advantages will undoubtedly be restricted to agriculture, much as they were during the classical genetic period.

The pessimists who fear the advance of man's knowledge in this area of molecular genetics are, I believe, unduly concerned. In any case, it is not possible to halt that advance, so interrelated are all aspects of biology. The millions we spend on a "cure for cancer" will, if productive, inevitably reveal some of the basic mechanisms for the integration of viral DNA into human chromosomes—mechanisms that relate directly to the procedures I have discussed. Certainly we must watch our step. But the question is: How can we best be on our guard—by restricting or by expanding our basic knowledge of life processes? Were we to opt for the meaner and more restrictive view, spending all our efforts in applied, goal-oriented research, we would increase the danger of moving in the wrong direction. And we would simultaneously cut off our view of new horizons that one day we may desire or need.

CIVILIZATION

Previous chapters have shown how science is probing secrets of earth, life, and health, and how many apparently unrelated sciences have collaborated in problem-solving. We turn now to those results of scientific research that we call civilization. Here, too, scientific discoveries are shown to be collaborative, the results of a continuous input from many different investigations and experiments by many different people. True, some aspects of civilization have caused men and women to doubt the benefits of research. But, as Dr. Asimov said in his Introduction, Einstein did not envision the mushroom cloud when he theorized that $e = mc^2$. In other words, unforeseen, unfortunate results of the application of knowledge produced by research into nature's laws does not mean that research should cease. Rather, it means that we need more research. As one prominent professor put it: "Science begets technical innovation, which frequently begets social change with feedback into the science base." Our next authors expand on that idea.

CHAPTER 10 HERMAN F. MARK

Polymeric Materials: Natural and Synthetic

> *Manifestly, not every finding leads straight to invention; but it is hard to think of major discoveries about nature, major advances in science, which have not had large and ramified practical consequences.*
>
> J. Robert Oppenheimer, in *Symposium on Basic Research*, 1959.

NATURE LEADS, TEACHES, AND IS OUTDONE

SINCE THE BEGINNING of his existence, man has relied strongly on the use of natural organic polymers for food, clothing, and shelter. (The word polymer is taken from the Greek and refers to molecules that consist of many (*polys*) and parts (*meros*); that is, they are very large.) When he ate meat, bread, fruit, or vegetables, and drank milk, he was feeding on proteins, starch, cellulose, and related polymeric materials. When he put on clothing made of fur, leather, wool, flax, or cotton, he used the same natural polymers. When he protected himself against wind and weather in tents and huts, he constructed these most primitive buildings from wood, bamboo, leaves, leather, and fabrics, which again all belong to the large family of organic polymers, adding to the above-mentioned types a few more species, such as lignin, resin, and bark.

But even later, when higher levels of civilization were reached, organic polymers were essential in both peace and war. All books in the famous library of ancient Alexandria consisted of either cellulose (paper) or protein (parchment), and they are made of these materials in all libraries of the world up to the present day. We would not have the Bible, the Talmud, the Koran, or any elaborate record of ancient history, were it not for the old scripts, which were put on wood, paper, or hides.

All transportation on land and sea throughout the centuries operated on wooden cars and ships, which were put in motion with the aid of ropes and sails made entirely of such cellulosics as flax, hemp, or cotton. The music of all stringed instruments is produced by the vibrations of proteinic fibers (catgut) and carried to the ear by the resonance of wooden, resin-treated boards. All famous paintings, together with many of the most valuable statues, consist of cellulose, lignin, and polymerized terpenes, in such materials as paper, canvas, wood, and paints. Bows and arrows are cellulose, lignin, resin, and protein; catapults and seige towers were made of wood and moved with ropes, and—until about 100 years ago—all discoveries of distant lands were made and all sea battles were fought with wooden ships that were maneuvered with the aid of cellulosic sails and ropes.

Although, in this way, natural organic polymers literally dominated the existence and welfare of all nations, virtually nothing was known about their chemical composition and structure. In each essential—food, clothing, transportation, communication, housing, and art—highly sophisticated craftsmanship developed, sparked by human intuition, creativity, zeal, and patience that led to accomplishments which will ever deserve the highest admiration of generations to come.

But even when, in the early decades of the last century, the chemistry of ordinary organic compounds became a respectable scientific discipline, the all-important helpers of mankind—proteins, cellulosics, starch, and wood—were not in the mainstream of basic chemical research.

Why?

Because somehow they did not seem attractive, at that time, for a truly scientific study, as they did not respond to the then-existing methods for isolation, purification, and analysis. The experimental backbone of organic chemistry in those days was dissolution, fractional precipitation, and crystallization; it worked then and it still works with all ordinary organic compounds, such as sugar, glycerol, fatty acids, alcohol, and gasoline, but it fails with wood, starch, wool, and silk. These materials cannot be crystallized from solution and cannot be distilled without decomposing.

This fundamental and embarrassing difference between the natural organic materials and the ordinary organic chemicals warned the chemists of the last century that there might be some essential and basic difference between these two classes of substances, and that one

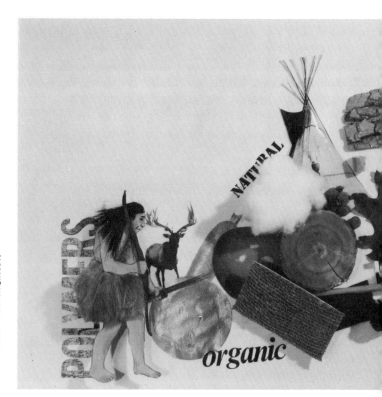

would have to apply special, new, and improved experimental methods to force the first class into the realm of truly scientific studies.

The breakthrough came in the early decades of the century, mainly through the adoption of physical methods, such as viscometry and osmometry (methods of determining the molecular weight of polymers by flow and diffusion), ultracentrifugation, ultramicroscopy, and

Since earliest times, man has used natural polymers.
Only recently has he discovered their chemical makeup.

X-ray diffraction. A decade of intense basic research on cellulose, proteins, rubber, and starch, carried out mainly in Europe, wound up with the following fundamental results:

1. All organic polymers consist of *very large molecules.* Whereas the molecular weights of such ordinary organic substances as vinegar, soap, gasoline, or sugar range from about 50 to 500, the molecular weights of the natural organic materials range from 50,000 to several millions, which earned for them the name "giant molecules," or "macromolecules."
2. Most of these molecules are made up of long, flexible chains, which are formed by the multifold repetition of a *base unit*, the so-called monomer (*monos* means "one" in Greek).
3. If a sample that consists of regularly built, flexible chains is exposed to stretching or rolling, the macromolecules are oriented and form thin, elongated bundles of high internal regularity. These bundles have been called crystallites. Depending upon the nature of the material and the severity of the treatment, more or less of a given sample undergoes crystallization, whereas the rest remains in an amorphous, or disordered, state so that any given fiber, film, rod, plate, or disc consists of two phases—amorphous and crystalline. This is shown diagramatically in Figure 1.
4. It was found that the crystalline domains contribute to strength, rigidity, high melting characteristics, and resistance against dissolution; the amorphous areas impart softness, elasticity, plasticity, and absorptivity.

As soon as the study of natural polymers had started to establish these ground rules, chemists were strongly tempted to try to synthesize equivalent compounds from simple, available, and inexpensive raw materials. The years from 1920 to 1940 saw ever-increasing successes that provided for more and cheaper basic building units—the synthesis of new monomers; that worked out efficient reactions to string them up into long chains—the mechanism of polymerization; that established quantitatively the molecular weight and molecular structure—polymer characterization; and that explored the influence of the structural details on the different ultimate properties—molecular engineering.

By learning originally from nature and following up on the established principles, scientists and engineers succeeded in producing a wide spectrum of polymeric materials which outdo their original native examples in many ways and, in most cases, are much more easily accessible and less expensive. All this gave a tremendous lift to the important industries of man-made fibers, films, rubbers, coatings, and adhesives and made everybody's life richer, safer, and more comfortable. Statistics show that in 1970 some four billion pounds of synthetic fibers were made and used in the United States; their total value was about $2 billion. During the same year, more than 10 billion pounds of plastics were produced and sold, representing a value of about $3 billion. Figures of the same order of magnitude hold for synthetic rubbers, coatings, and packaging materials. As a consequence, synthetic organic polymers have become a significant factor in the economy of all industrialized countries in the world.

DUAL HIGHWAY OF SYNTHESIS

Early forerunners of the work had found that certain by-products from the oil-refining industry contained double bonds and showed a tendency to "set" into amorphous, transparent masses under the influence of heat or light. (Bonds hold atoms together by an attractive force.) One of these by-products is styrene (C_8H_8), whose bonding is shown below:

$$\underset{H}{\overset{H}{\diagdown}}C = C\underset{C_6H_5}{\overset{H}{\diagup}}$$

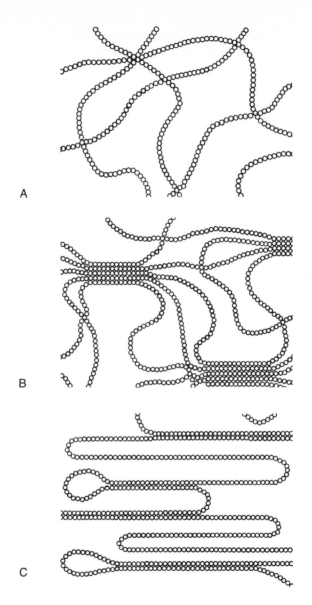

FIGURE 1 *Crystallization and disordered array of a polymeric molecule. A. Linear polymer chains in a random, disordered, or amorphous arrangement. B. Linear polymer chains that are partly aligned with each other, resulting in interchain crystallization. The rest are disordered. C. Linear polymer chains that show partly interchain crystallization, partly chain folding. There are no disordered arrays.*

Another is butadiene (C_4H_6):

$$\underset{H}{\overset{H}{\diagdown}}C = C - C = \underset{H}{\overset{H}{\diagup}}C$$

(with H atoms attached as shown)

In 1839, in the hands of the English scientist E. Simon, liquid styrene was transformed by heat into a colorless, brilliantly transparent, hard, and somewhat brittle resin. In 1887, O. Wallach, a German, found that another product of oil refining, liquid isoprene, would, if irradiated, form a colorless, soft, elastic material that had striking analogies to natural rubber. Others observed that, at elevated temperatures, small molecules with particularly reactive atomic arrangements—for instance, such groups as $-CH_2OH$(hydroxyl), $-COH$(aldehyde), $-COOH$ (carboxyl), or $-CH_2NH_2$ (amino)—would give hard, amorphous resins (when water was removed) that looked like wood or horn and, in many cases, could be neither fused nor dissolved. When heating phenol with aldehydes, the German scientist Adolf Baeyer found in 1872 that a clear, hard, and highly stable amorphous resin was formed and, as early as 1833, the French chemist J. Gay-Lussac obtained an amorphous, insoluble, resinous material by heating lactic acid. Toward the beginning of this century, an enterprising young Belgian chemist named Leo Baekeland even succeeded in starting the first commerical use of polycondensation products. They are still known today as Bakelite, a word coined from his name.

Such individual and scattered incidents of unwanted or wanted polymer formation became more and more frequent up to the middle 1920s, but it remained for Wallace Hume Carothers—the great pioneer of polymer chemistry, who carried out his most important work at the DuPont laboratories from 1928 to 1937—to establish a broad and clear helicopter view of the art and science needed to prepare synthetic polymers. Two main routes can be taken to arrive at chainlike (or linear) giant molecules: polymerization by addition and polymerization by condensation.

If styrene is heated or irradiated, its double bond "opens"

$$
\begin{array}{ccc}
\mathrm{H} \quad \mathrm{H} & & \mathrm{H} \quad \mathrm{H} \\
\mathrm{C} = \mathrm{C} & \longrightarrow & -\mathrm{C} - \mathrm{C} - \\
\mathrm{H} \quad \mathrm{C_6H_5} & & \mathrm{H} \quad \mathrm{C_6H_5}
\end{array}
$$

and two free valencies are formed. These react with other monomers

$$
\begin{array}{cccccccc}
& & & \mathrm{D} & & & \mathrm{S} & \\
\mathrm{H} & \mathrm{H} & \mathrm{H} \downarrow \mathrm{H} & & \mathrm{H} & \mathrm{H} \downarrow \mathrm{H} & & \mathrm{H} \\
-\mathrm{C}- & \mathrm{C}- +\mathrm{C} = \mathrm{C} & \longrightarrow & -\mathrm{C}- & \mathrm{C}-\mathrm{C}- & \mathrm{C}- + \ldots . \\
\mathrm{H} & \mathrm{C_6H_5} & \mathrm{H} \quad \mathrm{C_6H_5} & & \mathrm{H} & \mathrm{C_6H_5} \quad \mathrm{H} & \mathrm{C_6H_5}
\end{array}
$$

This simple addition of individual monomers to each other leads eventually to long-chain molecules. Each step of the chain growth involves the opening of a double bond (arrow D in formula 1) and the closing of a single bond (arrow S in formula 1). The first process requires about 20 kilocalories per mole less energy than the second delivers. Consequently, the chain growth is strongly *exothermic*, which is to say it "runs on its own power," represents an exothermic chain reaction, and "burns like a flame." Many hundred simple organic molecules can be induced to undergo addition-polymerization with the aid of heat, light, and a catalyst.

It was immensely important for the almost explosive development of the field that several of the molecules are readily available and inexpensive. Table I enumerates a few of them; their list can be lengthened substantially. Depending on the monomer used, additional polymers range from very soft and tacky elastomers (polybutadiene, polybutylacrylate) to very hard and strong plastics (polypropylene, polyacrylonitrile). Millions of pounds of them are produced and consumed every year as rubbers, plastics, films, coatings, and adhesives.

Two other petrochemical products figure in this picture. If adipic acid (AA) is heated with hexamethylene diamine (HMD),

TABLE I. *Important monomers used in addition polymerization*

Monomer	Formula	Made of	As polymer applied for	Price in 1970 in cents per pound
Ethylene	$H_2C = CH_2$	cracking of oil	garbage cans, bottles, toys, pipes, films	2-3
Propylene	$H_2C = CH - CH_3$	cracking of oil	same as ethylene; *also* carpets, upholstery	2-4
Vinylchloride	$H_2C = CHCl$	ethylene and chlorine	same as propylene	4-6
Butadiene	$H_2C = CH - CH = CH_2$	cracking of oil	mainly synthetic rubber	5-7
Styrene	$H_2C = CH - C_6H_5$	ethylene and benzene	same as ethylene	7-9
Acrylonitrile	$H_2C = CH - CN$	propylene, ammonia, and oxygen	synthetic fibers rubbers	10-12

$$\overset{\displaystyle C}{\downarrow} \qquad \overset{\displaystyle A}{\downarrow}$$
$$HOOC-(CH_2)_4\ COOH + H_2N-(CH_2)_6\ NH_2\ , \qquad (II)$$

the carboxyl group (arrow C in formula II) reacts with the amino group (arrow A in formula II) when water (H_2O) is removed to form

$$HOOC-(CH_2)_4\ CO-NH-(CH_2)_6\ NH_2 + H_2O. \qquad (III)$$

The product in formula III still has two reactive or functional groups left: a carboxyl group on its left end and an amino group at the right end. With these groups, it can react further with other monomer molecules and, eventually, form a long chain in which the residues of HMD and AA alternate regularly. Each addition of a monomer involves the formation and elimination of water, which requires energy. As a result, a polycondensation process represents an *endothermic* step reaction that does not "run on its own power" but requires continous heating if it is to be driven to higher conversions—that is, if one wants to form really long-chain molecules. (Because parts of the monomers −H and OH− are eliminated to form H_2O at each growth stop, these processes were called "polycondensation reactions" by Carothers.)

Again, several hundred monomers are available for polycondensation; a few particularly important ones are presented in Table II; their list could be prolonged at will.

The spectrum of useful products is here as broad as in polyaddition, and the two methods complement each other in providing an impressive variety of fibers—nylon, Fortrel, Orlon, Dynel; rubbers—Natsun, Neoprene, Hypalon; plastics —Polythene, Lustrex, Lexan, Zytel; coatings and adhesives.

FROM THE EGG TO THE CAKE

Discoveries and inventions deserve the attribute of "great" or "important" only if they make a significant and lasting impact on humanity. This impact may involve widening substantially the horizons of our understanding of nature—the universe, gravity, atomic structure, genetics—or it may provide new ways and means to make everybody's life longer, safer, and better. In both cases, there is a lapse between the first decisive steps and the final and universal acceptance of their consequences and their significance. After Columbus had set foot on a Carribean island in 1492, it took many years until the true significance of the feat was generally realized. When Neil

TABLE II. *Important Monomers for Condensation Polymerization*

Monomer	Formula	Made of	Used for	Price in 1970 in cents per pound
Ethyleneglycol	$HOCH_2—CH_2OH$	ethylene and oxygen	fibers, films, plastics	6-8
Adipic acid	$HOOC—(CH_2)_4—COOH$	benzene plus H_2 and O_2	nylon fibers, plastics	18-20
Hexamethylene diamine	$H_2N—(CH_2)_6—NH_2$	benzene and NH_3	nylon fibers	26-28
Caprolactam	$\begin{matrix} CH_2—CH_2—NH \\ \mid \qquad\qquad CO \\ CH_2—CH_2—CH_2 \end{matrix}$	phenol, H_2 and NH_2OH	nylon fibers, plastics	18-20
Terephthalic acid	$HOOC—\langle\bigcirc\rangle—COOH$	p-xylene and O_2	fibers, films, plastics	13-15

Armstrong made "the small step for man" down to the surface of the moon, all parts of the civilized world immediately saw on their TV screens this "giant leap for mankind." A similar shortening of the incubation period of discoveries and inventions has occurred in the field of polymer science and technology. In 1872, Adolf Baeyer had on his laboratory bench the first phenolic resin, but not until 1905 did Leo Baekeland start its larger-scale preparation. The product did not reach the market until 1913. In 1870, Baeyer also became the first to synthesize the plant dye indigo, using a method that was not commercialized by the BASF, a German chemical plant, until 20 years later.

Things move more rapidly today because the path from the initial spark to the final fire is now thoroughly paved and the individual steps are well organized. The primary ingredients for success are: creative imagination, careful experimentation, and critical evaluation of the results. This far, everything happens in the brain, on a piece of paper, and in the laboratory; expenses are moderate and progress is usually rapid. But then comes the scaling up and the development of a safe process and a salable product. This needs much money, many people, and a lot of patience.

However, if a large and profitable market can be seen in the early stages, even the step from the laboratory to the 50-million-pounds-a-year plant can be accelerated miraculously. In England, in 1933, E.W. Fawcett and his associates showed to the management of the Imperial Chemical Industries the first few grams of polyethylene in the form of a white powder. In spite of the unusually high pressure that was needed to polymerize ethylene—15,000 psi—commerical production began in 1937. Later—in 1954 and 1955—when the German Karl Ziegler and the Italian Giulio Natta invented a low-pressure synthesis of polyethylene and polypropylene, it took not more than four or five years to establish commercial production. At the same time, the Phillips Petroleum Company in the United States got its own low-pressure polyethylene process under way in a similarly short period. Neoprene and nylon were the results of careful, fundamental studies of addtion and condensation polymerization. It was obvious to Carothers that interesting synthetics would certainly be found somewhere in their domain, but his approach remained basic and systematic at all times.

When, in 1933, he and J. Hill had in their test tube a typically fiber-forming polyamide, development work was pressed by DuPont with great energy, and the first nylon plant started production in Seaford, Delaware, in 1937. Within the same short period, Neoprene developed from fundamental experiments in Carother's laboratory to the first commercial synthetic rubber produced in the United States that displayed superior resistance to gasoline, oil, light, and oxygen.

Another impressive and intriguing success story that is intimately connected with the use of polymers is Edwin Land's invention of Polaroid sheets and his subsequent untiring efforts to give instant photography to the public. Black-and-white photographs and the large-scale production of polarizing sunglasses and windows resulted in a few years. The development of instant photography took the longest time, because it was not possible to utilize existing polymers as carriers and barriers for developer and fixative. Instead, new macromolecules had to be synthesized in order to be tailored for the specific job they had to perform in the multilayer film. Yet, five or six years after the original idea of instant photography was born, the first black-and-white Polaroid camera and film were on the market. Additional difficult and painstaking basic research was necessary before adequate dyestuffs and barriers for the much more complex process of instant color photography were found. But again, systematic and fundamental research paved the way for industrial production; by the mid-1960s, Polaroid color film was given to the public as another miracle of our day, and Edwin Land emerged as the wizard of modern picture making only about 130 years after Daguerre had started it all in France about 1830.

WHAT NEXT?

Science and technology are in an enviable position: they not only tell us what we have but they also invite us to take reasonably legitimate peeks into the not-too-distant future. The laws of nature can be represented either as algebraic equations or as their geometric counterparts in the form of curves. An equation always stimulates a scientist to speculate what he would get if he made slight—and therefore probably realistic—changes in the numerical parameters of his equation, and a curve tempts him even more to continue it a little bit beyond its actually established limit, using the slope and curvature of its last elements. Since Leibniz and Newton, science has successfully used these recipes, which, in fact, are nothing but the principle of the calculus and of variation theory.

Technology is in a similar situation, only it operates with materials and methods instead of with equations and curves. But here, also, the engineer's fantasy is strongly stimulated by assuming that small—and therefore probable—improvements of his materials and methods will let him reach a new level of performance. Later, when his assumptions have been verified experimentally, he has attained a new and reliable platform for further progress. Actually, this is nothing but modern operational analysis. Let us now apply to our polymers these well-established principles of forecasting and see where we get with them.

HIGHER SPEED, BETTER CONTROL

Commercial, large-scale, polymerization processes are presently carried out at rates that are directed essentially by the need for material and heat transfer. Laboratory chemists are able to carry out the processes up to 10 times faster in smaller scale. Similar speeds in commercial operation by improved reactor designs are even now in the planning stage, and there can be little doubt that these improvements will be in operation in two or three years. In most present commercial polymerization processes, the molecules grow with little or no control; the result is like a poorly managed forest—some trees are high, others are low, and there is a lot of dead wood. The above-mentioned modern reactor designs permit not only accelerated rates but also improved control of the chain lengths and configurations of molecules. Result: more uniform polymers in shorter lengths of time, which means better products at lower cost for all uses.

HARDER, STRONGER, LESS CORROSIVE

There exist today organic fibers that have a modulus, or rigidity, and a strength approaching that of soft steel (20 million psi and 500,000 psi, respectively). On top of that, they have a specific gravity of only 1.3 as compared with 7.5 for steel. Their strength-to-weight ratio is therefore superior to that of steel. If they are used to reinforce thermosetting resin compositions (those that harden when heated), one arrives at plates, rods, and pipes that are as strong as metals but are less corrosive. Tanks and barrels for milk, gasoline, and beer are made of such compositions, as are pipelines for crude oil, water, and natural gas. Even more popular is their use for the construction of sailboats and motor yachts. Improvements in properties that have already been established in the laboratory and lowering of production costs that come with larger-scale use allow us to predict that in a few years even very large ships, most pipelines, and, to a considerable extent, the doors, hoods, lids, and fenders of cars and the entire fuselage of planes will be made of fiber-reinforced, thermosetting, organic resins. Architects have shown intensive interest in these lightweight materials, which have exceptional resistance to mechanical and chemical changes; they visualize using the resins to build gigantic domes, halls, and towers. As a result, buildings in the 1980s will dwarf present constructions in size, style, and elegance.

LOWER CREEP, LESS FATIGUE

Improving chain molecules in respect to their length and uniformity reduces "long-time creep," which means a slow deformation with time, and improves resistance against fatigue from periods of a few months to many years. Hence, high-modulus, high-strength organic fibers with melting points at about 500° C are now being tested for elevator cables, ski lifts, and cable cars. A few lightweight suspension bridges have already been designed with cables consisting of high-modulus fibers, and it is quite probable that the largest bridges of the future will not be made of steel and concrete but of aromatic nylon cables and fiber-reinforced, thermosetting polyester or polyamide compositions.

SOUNDER AND SAFER

Today organic polymers are enlisted by scientists and

engineers in the fight against the pollution of our world. In one approach, graphite fibers are used for screens that remove dust and acids from the waste gases of industrial smoke-stacks in refineries, in steel and paper mills, and in electric power plants. These filaments and wires are made by the pyrolysis and graphitization of organic fibers. In the former process, the fibers are treated at 1000°C in the absence of air; in the latter process at 2000°C. These techniques make the fibers as rigid and strong as steel. They can be used at temperatures up to several thousand degrees and are much lighter and more corrosion-resistant than are any competitive metals.

The purification and eventual recovery of aqueous wastes operate in three stages. First, the coarsest impurities—suspended solid particles—are removed by sedimentation or flotation. Organic polymers are used in this step as foaming and flocculating agents. Next, finer particles of colloidal dimensions are eliminated by ultrafiltration (UF) and, finally, dissolved organic and inorganic substances—salts, acids, bases, etc.—are removed and recovered by reverse osmosis (RO). Considerable effort, sponsored by both government and industry, is now concentrated on preparing new polymers for the manufacture of improved membranes for UF and RO, and a state is already reached at which commercial desalination of sea water and complete purification of city sewer effluents, together with recovery of valuable organic materials, are matters of only a few years of additional development.

Great progress also has been made recently in rendering organic polymers flame-retardant and, in some instances, completely nonflammable. This progress has led to safer toys, clothing, curtains, floor coverings, insulating walls, roofs, and partitions, and has also given valuable hints and clues for further improvements.

THE MAN-MADE MAN

Synthetic plastics are stronger and tougher than bones, synthetic rubbers more resilient than muscles, and synthetic membranes more durable than skin. Yet all presently available materials, in spite of their superior mechanical behavior, can be used only with vital restrictions and great caution, because they are not compatible with the processes and functions of life; they coagulate blood, create irritation, and, eventually, are rejected by the body.

The use of hard plastics in the replacement of limbs and particularly in dentistry is already well developed. Additional and substantial progress will be made with the introduction of coatings that bind with the tissues of the body without irritation; by adhesives and cements that set in the body in a very short time (less than one second) to form tough and resilient ligaments; and by strong, rigid, and unbreakable foams that permit artificial limbs to be made infinitely lighter than those used today (Figure 2).

Soft organic polymers in the form of membranes, fibers, and tubes are already used to replace blood vessels, tendons, muscles, and skin. Difficulties that still must be overcome are the insufficient binding capacity of the synthetic part with the natural organ, the irritation some synthetics produce and, specifically, their tendency to coagulate the blood, which may cause dangerous blood clots to form.

These flaws can be eliminated, and much research to that end is being conducted cooperatively by medical and chemical experts. Systematic progress is, therefore, only a question of time, and an unexpected breakthrough could come at any moment.

CONCLUSION

The next chapter discusses some instruments that have changed the face of our world—instruments that have developed as a result of man's need to know. Here, too, basic research in polymer chemistry has made vital contributions. For instance, plastic monofilaments about $\frac{1}{32}$ of an inch thick can conduct light around curves and in spiral paths. This finding has led to a number of sophis-

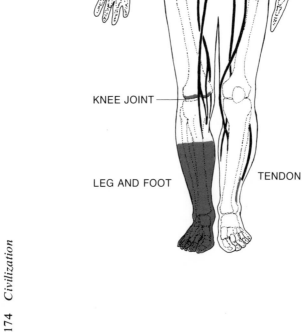

CORNEA

EAR   NOSE

TEETH

BONE
REPLACEMENT

HEART
VALVE

BLOOD
VESSELS

HIP JOINT

KNEE JOINT

LEG AND FOOT   TENDON

ticated diagnostic techniques in medicine. All tanks, planes, and submarines are equipped with unbreakable plastic lenses and prisms. Soon, all contact lenses will be made of polyacrylics; automobile headlights and reflectors already are made of plastics. Plastics have replaced glass in all sunglasses, most heavy spectacle lenses, magnifying glasses, and many inexpensive microscopes and telescopes. Thus, in a very real way, man's increasing knowledge of the nature of organic materials has given him an improved life style, and has opened broad new pathways for the future.

FIGURE 2   *Many plastics are accepted by the body's own tissues. They also show great chemical, physical, and biochemical stability. These advantages have given rise to new and dramatic surgical techniques.*

A new cornea can be made from a thin layer of cartilage taken from the patient's body and containing a hole filled with clear acrylic plastic that acts as a lens. Teeth are being manufactured from acrylic plastic, and many artificial limbs are now made of strong, light, epoxy foam. The plastic surgeon replaces damaged tissues with silicone material; in this way, even artificial ears and noses can be fashioned. Plastic tubing, often made of dacron, can function as blood vessels or tendons. Injured hip and knee joints can be repaired with polyethylene. Artificial heart valves of various polyesters also have proved to be remarkably functional.

The Better to See...

LITTLE RED RIDING HOOD: *"But Grandma, what big eyes you have!"*
WOLF: *"The better to see you with, my dear!"*
From: *"Little Red Riding Hood"*

THREE OPTICAL INSTRUMENTS—the telescope, the microscope, and the spectroscope—have revolutionized man's view of nature. Not only have they shown us familiar objects more clearly, but they also have revealed phenomena so unsuspected that man has been forced to revise age-old ideas drastically. Moreover, each of the three is undergoing rapid transformation in our own times, so that an end to their revelations is not yet in sight.

They began, really, with the work of the unknown artisans who, in the thirteenth century, created spectacle lenses. Galileo selected certain of those lenses and ultimately made a telescope good enough to use in his studies of the night sky. With only this primitive equipment, he discovered the mountains on the moon (Figure 1), the phases of Venus, and, most startling, the satellites of Jupiter. As a result, the old view that the earth was the center of the universe was finally shattered because, unquestionably, certain heavenly bodies revolved around something other than the earth.

The microscope was invented about the same time as the telescope (Figure 2). As early as 1590, the Dutch optician Zacharias Janssen made a compound microscope

FIGURE 1 *Galileo's primitive telescope showed that the moon was mountainous and pitted, instead of smooth and perfect, as Aristotle had described it. This has been amply supported by modern telescopes and, of course, by space travel. The modern photograph, taken by a telescope at the Lick Observatory in California, shows the full moon—14 days old.*

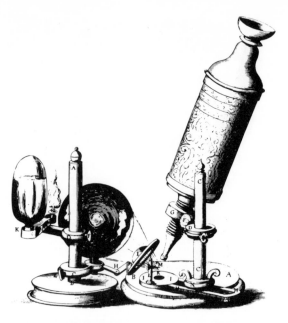

FIGURE 2 *This engraving, cut by the eminent British scientist and philosopher Robert Hooke, was published in 1665 in the first treatise devoted solely to microscopy. It shows Hooke's improved compound microscope with its hemispherical lens objective and plano-convex lens eyepiece. A spherical condenser concentrated rays from a lamp on the specimen.*

with an eyepiece made of a convex lens. The microscope led to discoveries in the latter half of the seventeenth century by the Dutch naturalist Anton van Leeuwenhoek. He used a simple one-lens microscope, which really was only a high-powered magnifying glass mounted on a stand, and with this simple apparatus, discovered life on a microscopic scale, including spermatozoa and bacteria. His findings were to revolutionize the thinking of the period.

In an article in *Science,* Joseph Weizenbaum noted that when the microscope was invented "the dominant commonsense theory of disease was fundamentally that disease was a punishment visited upon an individual by God. . . . The microscope enabled man to see microorganisms and thus paved the way for the germ theory of disease. The enormously surprising discovery of extremely small living organisms also induced the idea of a continuous chain of life which, in turn, was a necessary intellectual precondition for the emergence of Darwinism. Both the germ theory of disease and the theory of evolution profoundly altered man's conception of his contract with God and consequently his self-image.... I do not say that the microscope alone was responsible for the enormous social changes that followed its invention. Only that it made possible the kind of paradigm shift, even on the commonsense level, without which these changes might have been impossible."

The next revolution—one that is still going on—came with the spectroscope. Isaac Newton had used a prism to disperse white light into its constituent colors in 1664, but real spectroscopes were not constructed until the nineteenth century, when it was found that different colors in the spectrum correspond to different lengths of the light waves. By 1861, Robert Bunsen and Gustav Kirchhoff, in Germany, had demonstrated that wave lengths, in turn, corresponded uniquely to definite chemical elements. As a result, several new chemical elements, such as rubidium, cesium, and subsequently helium, were identified through their distinctive spectra. For the first time, it became thinkable and, indeed, possible, to learn something about the actual chemical composition of the sun, stars, and planets.

In the twentieth century, when good photographic spectrographs were connected to large telescopes, men could see that spectral lines from distant galaxies were quite generally shifted to the red end of their normal wave lengths. This indicated that the galaxies are receding from us. Edwin Hubble, in the United States, found that the more distant galaxies are receding fastest, and thus was born the concept of an expanding universe born in a cataclysmic explosion some few billion years ago, as was discussed by Dr. Hoyle in the first chapter of this volume.

But the most revolutionary consequences came from

the attempts to understand why particular atoms emit their characteristic wave lengths. By the latter part of the nineteenth century, it was known that atoms must be several times smaller than the wave length of visible light. It was also known that light waves diffract around objects as small as atoms and so cannot form images of them. As a result, we can never hope to see the shape of atoms by looking at them with light. However, studies of the spectrum of light emitted by atoms led to the discovery of quantum mechanics. In this theory, electrons behave in some ways like waves and in other ways like particles. Moreover, quantum mechanics has explained the way in which light is absorbed and produced. This knowledge played an essential part in the invention of lasers, which I discuss later in this chapter.

TELESCOPES TODAY

Although spacecraft have reached the moon and the nearer planets and so made possible close inspection of these bodies, the stars and nebulae are vastly more distant, and remain as inaccessible as ever. Thus, much of our information about astronomical objects must come through telescopes of various kinds linked with other instruments, such as spectroscopes.

As we have tried to look for fainter and more distant objects, telescopes have been made larger. The bigger the area of the main mirror, the more light it can gather. To record the image, we have long since replaced the astronomer's eye with a photographic plate, which can total all the faint light that is received from stars during an exposure of some hours' duration. More recently, improved television camera tubes, considerably faster than photographic materials, have been applied to astronomy.

Some earth telescopes developed in the last generation use radio waves rather than light. These radio telescopes are large, sometimes several hundred feet in diameter, and when coupled to sensitive receivers can detect very faint and distant objects. But radio waves are typically about a million times larger than light waves. Big as the radio telescopes are, their ability to resolve neighboring objects is much poorer than that of optical telescopes. Better resolution can be attained by using arrays of two or more antennas, whose signals are sent to a central receiver to be compared and processed. An array of antennas can be made to resolve details as fine as those that could be discerned by a single huge telescope. To push this method to the limit, the antennas can be located on different continents, thousands of miles apart. Rather than transmit the signals to a common processer, which would be extremely difficult to do over such large distances without distortion, the signals are recorded at the two stations, along with accurate timing markers from an atomic clock. Then the records can be synchronized precisely and analyzed by a computer. In this way, radio astronomers are now able to achieve the resolution of a telescope that is, in effect, as large as the earth!

Sensitive detectors, developed in recent years, make it possible to observe the sky in infrared and ultraviolet wave lengths. Most of these regions cannot be observed from the ground, although some tantalizing glimpses can be obtained from high mountain tops or from high-flying airplanes, balloons, or rockets. Actually, the first astronomical sources of X-rays were discovered by rockets, which briefly emerged from the atmosphere and then returned to earth.

Now, however, it has become possible to place an observatory outside the earth's atmosphere. The problems are formidable, because a precision instrument of this kind must be made to operate for long periods without repairs. The first such instrument already exists in the Orbiting Astronomical Observatory, OAO-2, which was launched in December 1968. This 4,200-pound satellite, built at the Grumman Aerospace Corporation, contains eleven scientific instruments, including telescopes equipped with television cameras sensitive to wave lengths as short as 1,000 Angstroms and with spectrometers for a wide range of ultraviolet wave lengths.

(One Angstrom unit, which is a measure used to express the length of light waves, is one ten-thousandth of a micron or one hundred-millionth of a centimeter.) The telescopes can be pointed in any desired direction with high precision, and respond to commands transmitted from a control station on earth.

OAO-2 has provided the information for a map of the whole sky as seen in ultraviolet light. A number of stars turn out to be much fainter in the ultraviolet than was expected from their appearance in earth-bound telescopes, whereas about 20 per cent of the objects seen near the plane of our Galaxy are so faint in the visible region that they had not been observed before (Figure 3).

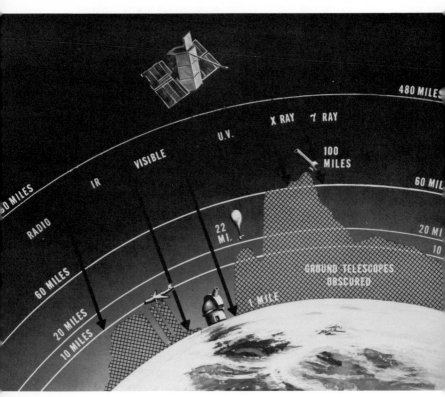

FIGURE 3 *This shows the way in which the Orbiting Astronomical Observatory has expanded the spectra available for astronomy. (IR = infrared; U.V. = ultraviolet)*

Now an even more sophisticated spacecraft, OAO-3—called "Copernicus"—is making telescopic observations for the first time in history, in both the ultraviolet and X-ray wave lengths, with a precision much greater than that of OAO-2.

With these radically new kinds of telescopes, astronomy has become one of the most exciting fields in science. One of the new discoveries is the possible existence of objects so dense that nothing, not even light, can escape. These are called, appropriately enough, black holes. Although black holes themselves cannot emit anything to signal their presence, they may be detected by their effects on neighboring regions, including the production of X-rays as neighboring matter is drawn into the hole. These effects can be observed by X-ray telescopes outside the earth's atmosphere.

Perhaps new laws of physics may be discovered by studying such celestial objects. Certainly, some astronomical objects appear to be emitting such enormous amounts of energy that it is difficult to find an explanation based on known physical processes.

MICROSCOPES TODAY

Enormous numbers of microscopes are in common use today, not only in laboratories and doctors' offices, but perhaps even more in factories that make such tiny devices as transistors. Microscopes of this sort represent a technology that is now old and well developed, but even here spectacular advances are occurring. One of these was the introduction of phase contrast by the Dutch physicist Frits Zernike in the 1930s. With phase-contrast techniques, a nearly transparent object can be distinguished from a transparent background by its slightly different refractive index. The method is particularly useful for studying cells and other biological specimens, which can be examined under a phase-contrast microscope without being stained.

But whatever the refinements in the design of light microscopes, they are always limited by the wave nature of light, which prevents them from showing clearly any object or detail much smaller than the wave length. The really dramatic advances in microscopes came when electrons were substituted for light. By the 1920s, it was known that electrons behave like tiny charged particles that can be deflected by electric or magnetic fields. But it really was not until the wave nature of electrons was discovered that anyone thought of trying to make lenses that would focus electrons.

In an electron microscope, a beam of electrons plays the part of the light rays in an ordinary microscope. The electrons can, if they are fast enough, pass through thin objects. They then are either partly transmitted or are absorbed, depending on the opacity of the various parts of the specimen. Magnetic coils serve as lenses to image the electrons onto a fluorescent screen or a photographic film. The electrons must be accelerated to fifty thousand up to a million volts before some of them can penetrate the specimen and produce the image.

These advanced microscopes have revealed many details of the structure of tiny organisms. For instance, they have clearly imaged a virus, including its head and tail (Figure 4). For these microscope studies, H. Fernández-Morán, at the University of Chicago, has developed a diamond knife, or ultramicrotome, which can cut specimens as thin as 50 Angstrom units—small enough to permit doing chemistry by cutting. For example, a large molecule—of starch, let us say—can be cut in such a way that it becomes sugar. The electron beam itself has been used to reduce a page of printed material to an almost invisible dot. Letters on the page are only about one hundred atoms high and, at this reduction, the whole collection of the Library of Congress could be contained on one sheet of eight-by-ten-inch paper. Of course, it would have to be read through an electron microscope, but printed matter is increasing at an enormous rate. One day, this reduction technique might be needed to store acres of books, documents, and other miscellaneous papers.

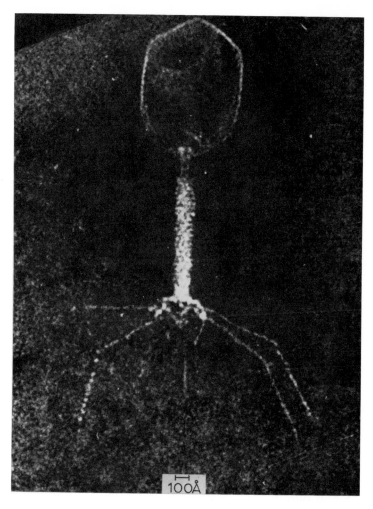

FIGURE 4 *High-resolution electron micrograph of a bacterial virus shows the structure of the head and tail. The picture was taken at X 300,000 magnification.*

One goal of the microscopist is to see the individual atoms that make up molecules. This would be especially valuable for complex molecules, whose structures are still unknown. This now is beyond the capability of even the most powerful electron microscopes, but progress is encouraging. For instance, the two-stranded structure of

DNA molecules has been observed in an electron microscope (see Chapter 9).

Individual atoms on surfaces can be seen in a different kind of instrument—the field ion microscope. An electric field is applied to the surface layer on a fine needle point of some metal, such as tungsten. The field is so strong that it can ionize atoms, pull them away from the surface of the specimen, and accelerate them in such a way that they produce bright spots when they strike a phosphor screen or a photographic plate. The pattern of these spots is an enormously enlarged image of the tip of the needle point (Figure 5).

Less extreme resolution but remarkable depth of field and clarity of the image are obtained in scanning electron microscopes. In these, the electron beam is focused to a single tiny spot, which is then scanned across the specimen. The scattered electron current—or, alternatively, the X-rays produced at the specimen—is detected and used to control the brightness of a spot on a picture tube, which is scanned in synchrony with the electron beam. If X-rays are used, they can be filtered so that the detector responds only to certain kinds of atoms. Thus the microscope can show the composition, as well as the location, of the various constituents of the specimen (Figure 6).

SPECTROGRAPHS AND LASERS

Spectrographs are essential tools in astronomical research, but they are also widely used today in physics, chemistry, biology, and many other branches of science and technology. Here, too, the wave-length range now extends far beyond the visible into the infrared and ultraviolet regions. Chemists and metallurgists utilize spectrographs to analyze all kinds of materials, and to detect traces of impurities. For instance, not long ago, there was a sudden fear that some fish might have become contaminated by mercury from industrial wastes and so were unsafe to eat. Sensitive spectrographic techniques were put to work, and it was found that the fish were indeed inedi-

FIGURE 5 *Field-ion micrograph of a hemispherical tungsten tip with a radius of approximately 300 Angstroms. The average magnification of this print is X 1,700,000. Each single dot is a tungsten atom; each ring represents the edge of a crystal facet.*

ble. Characteristic spectra also can betray the presence of contaminant gases in the atmosphere and can be used to monitor the emissions of automobiles. Physicists use spectrographs to study the composition of matter, the structure of molecules and solids, and the forces holding them together.

The growing understanding of how light is emitted by

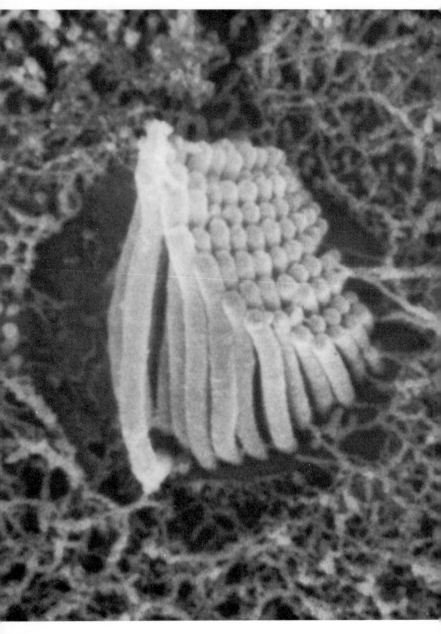

FIGURE 6 *Scanning electron micrograph of hair cells in the vestibular receptor of a mudpuppy—a large salamander. Its magnification is X 30,000.*

atoms and molecules has also led to the totally new kinds of light sources known as lasers. The word is an acronym for light amplification by stimulated emission of radiation. Instead of the random jumble of many-colored lights that any conventional lamp emits, a laser produces a narrow, sharply directional beam that has a pure, single color. And those beams have power densities many millions of times greater than light beams on the sun's surface! There is now a large family of different kinds of lasers, but all of them use stimulated emission of light from an assemblage of atoms that have been temporarily excited so that they have some stored energy which can be released under the stimulus of a suitable light wave. Nearly all of them also use the structure that Charles Townes and I suggested in 1958, in which the excited atoms are arranged in a long, pencil-like column with mirrors facing each other at the ends. One of the end mirrors is partly transparent, so that when a strong light wave grows as the excited atoms in the column are stimulated to emit, part of it emerges as the laser's output beam.

Lasers range in size from tiny, transistor-like, semiconductor lasers to giant gas lasers hundreds of feet long. Probably the most widely used is a gas-discharge laser, rather like a small neon sign, that uses a mixture of helium and neon gases to produce a beam of red light. The power output of such a laser is small, but its highly directional beam provides a nearly ideal straight line to guide machines for drilling tunnels and for laying sewer pipes. Other industrial uses of lasers include metal work, welding, cutting, and drilling.

Low-power lasers can be extremely useful in communications. Light sends out millions of times more waves per second than does a TV signal, so correspondingly more information can be carried by a light beam. In principle, one light beam could carry as much information as all of the radio channels combined, and it could be directed sharply to a distant receiver. These properties will make laser communication useful in outer space, where television signals could be sent back from spacecraft as far away as the planet Mars. But on earth, rain, snow, or fog can interfere with laser-light transmission. In the future, laser beams might be sent through straight pipes, which would protect them from the weather. Thin glass fibers can also carry light signals over distances of a mile or more. In large cities, particularly, light carried through fibers may soon become an important means of communication.

Other small lasers can produce short light pulses that range in power from a few hundred to a few thousand watts. Although these pulses may last for just a small fraction of a second, they can be focused to an extremely bright spot, which is intense enough to vaporize a small amount of any substance. Lasers of this sort are being used to drill tiny holes in diamonds and in sapphires for watch bearings.

Moreover, a laser beam can be focused to produce intense heat at a place which can be seen but not touched, such as the retina of the human eye. Other methods of repairing detached retinas have not been entirely satisfactory. However, by using the laser method, doctors can make small lesions on the retina that help to hold it firmly against the supporting back of the eye. They can also use a laser to seal leaky blood vessels in the eye—a relatively common result of diabetes. Other applications to surgery seem possible, although they are still only speculative. For instance, a laser knife could be used for surgery on organs that have many blood vessels, because it can seal their ends as it cuts. Experiments have also shown that lasers may one day be used extensively for removing "birthmarks," warts, and even tattoos!

Lasers that produce continuous beams at a power level of several hundred watts are being applied broadly by industry. They can be controlled by a computer, as in a machine that cuts patterns from cloth.

Lasers offer another prospect that is even more exciting. When a high-power laser is focused on a small spot, the light is so intense that one day it may be possible to heat pellets of heavy hydrogen sufficiently to induce a thermonuclear reaction. If this can be done in a controlled way, unlimited supplies of energy, nearly free of radioactive waste products, will become available. However, a process of that kind would call for temperatures of millions of degrees, and the heat must be applied quickly enough to prevent the pellet from flying apart before the thermonuclear reaction can take place. The task is difficult, and really large, complicated lasers will be needed to give carefully shaped pulses of some trillions of watts for a few trillionths of a second (see Chapter 14). Success is not certain, but the researchers are optimistic and the need for energy sources is great and growing, as pointed out by Dr. McKelvey and Dr. Starr in Chapters 13 and 14.

Even a small laser, such as the helium-neon type, is sufficient for holography, one of the prettiest applications of lasers. This novel method of photography, invented in England by Denis Gabor in 1948 but made practical by the advent of lasers, can provide a truly three-dimensional image. A typical hologram looks like a half-exposed sheet of film, on which appears only a nearly uniform blur. It is an extremely fine, complex pattern of lines and spaces too fine for the eye to see directly, but when it is illuminated by a laser, an image is formed. Then the hologram behaves like a magic window through which you can see the object. You look through the hologram with both eyes, and the object appears to be floating in space. If you move your head, you see a different view, and may be able to look behind one object to see another, just as you might if you were looking through a window.

To make a hologram, light from a laser is used to illuminate whatever is to be photographed. Some of the laser light is split off by a partial mirror and allowed to fall directly on the photographic film, along with the light scattered by the surface of the object. On the film, the scattered light and the reference beam combine to produce a complex interference pattern that is the hologram.

Many applications have been suggested for holography, such as preserving records of valuable statues in case the original should be damaged. Holograms can be used to store information for computer memories, in a form that may have advantages for various kinds of information retrieval.

An intriguing prospect for the future may be an extension of holography to the X-ray region. When X-ray lasers are achieved, they may make possible holograms from which enlarged images can be reconstructed with visible light. Such holograms may very possibly provide the means for seeing the individual atoms in molecules.

CONCLUSIONS

These three optical instruments—telescope, microscope, and spectroscope—have made it possible for people to see more and to understand what they see. The things revealed by these instruments have shaken men's views of themselves and of the universe. The new knowledge has led to the creation of dramatic new tools, such as lasers.

As these instruments become more powerful, drawing on discoveries of science and technology, their revelations continue to surprise us. Most of the discoveries serve to broaden our understanding and appreciation of the basic laws of physics. But somewhere, as we continue to explore, there is always a chance that we will find a phenomenon that will force us to recognize new and even more fundamental principles and concepts. Will the telescope show us an unknown source of energy that we can learn to harness? Will we learn enough about the structure of biological molecules so that we can practice genetic engineering? The only thing certain is that the future holds promises for all mankind.

CHAPTER 12 MARK KAC

Will Computers
Replace Humans?

*I should say that most of the harm computers can po-
tentially entrain is much more a function of properties
people attribute to computers than of what a computer can or cannot actually be made to do.*
Joseph Weizenbaum, in *Science*, May 12, 1972

THE COMPUTER, one can safely predict, will be adjudged
to have been the ultimate technological symbol of our
century. Although not nearly as common as a car or a
TV set (the runners–up in the race for the ultimate technological symbol), it has affected our views, our atti-
tudes, and our outlook in more subtle and disquieting
ways.

We find it difficult to accommodate ourselves to the
computer because, on the one hand, it has proved to be
an indispensible tool in such great adventures as that of
putting man on the moon and, on the other hand, it threatens our fundamental rights by its ability to keep and to
hold ready for instant recall prodigious amounts of information about our activities, our habits, and even our
beliefs.

Every gasp of admiration for the remarkable feats it
can perform is matched by a groan of pain caused by its
apparent inability to shut off a stream of dunning letters
long after the bill has been paid.

Never has there been an instrument of such capacity

for the useful and good and at the same time of such potential for mischief and even evil. And what makes our coming to terms with the computer so difficult and so frustrating is that we find ourselves face to face with a being seemingly capable of some imitation of the highest of human acts, namely that of thought, while at the same time totally devoid of all human qualities.

"To err is unlikely, to forgive is unnecessary" is the computer's version of the old proverb that "to err is human, to forgive divine" (page 206). This is frightening, especially if one thinks of the antics of HAL, the deranged computer of "2001."

We might be less frightened if we knew a little more about modern computers: what they are, what they can do and, perhaps most importantly, what they cannot do.

Computers of one kind or another have been with us for a long time. The abacus came from antiquity and is still in use. The slide rule is probably close to being two hundred years old and today is no less indispensable to scientists and engineers than it ever was. The adding machine has been in use for well over half a century and such "business machines" as sorters and collators still account for a significant proportion of IBM rentals.

Throughout history, man has repeatedly tried to relegate the drudgery of computing to a machine and to increase the speed and accuracy of routine calculations (Figure 1). Perhaps the most remarkable of all such efforts was the work by Charles Babbage (1792-1881), because in many ways it foreshadowed the advent of modern electronic computers. Babbage's inspired and clever "calculating engines" came a bit too soon, for neither the minds of his contemporaries nor the technology of his day were quite ready for them.

The present-day, all-purpose digital electronic computer (to give it its rightful name) is a product of the unlikely union of advanced electronic technology with some quite abstract developments in mathematical logic.

The spiritual father of the modern computer is the late

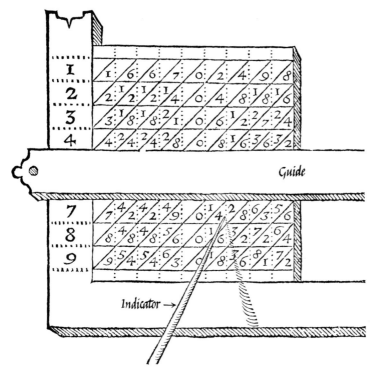

FIGURE 1 *"Napier's bones" constituted one of the earliest forms of mechanized calculation. With these, John Napier (1550-1617) tried to simplify calculating, and coined the word* Rabdologia *(studies with a rod or stick) for numeration by little rods. Eventually, the rods were mounted on a cylinder and an arrangement of these in a box constituted a calculator. (Mechanized addition and subtraction were introduced in 1642 by Pascal, then 19 years old.)*

A. M. Turing, a highly gifted and original English mathematician, who in 1936 conceived of a "universal machine" that now bears his name.

The machine consists of an infinite tape divided into identical squares. Each square can either be blank [which is denoted symbolically by (*)] or have a vertical bar (|) in it. There is also a movable scanning square that can be moved either one step to the right or one step to the left.

The only operations allowed are:

L : move the scanning square one step to the left
R : move the scanning square one step to the right
* : erase the bar (|) (if it is there)
| : print the bar (|) (if it isn't there)

Instructions to the machine are in the form

$$5 : * \mid 7 ,$$

which is read as follows:

> Instruction 5 : If scanning square is over a blank
> square, print | and see instruction 7,
> Instruction 7 may read as follows:

$$7 : \mid R \; 7 ,$$

i.e., if the scanning square has a vertical bar in it, move it to the right and repeat instruction 7. However, if, after following instruction 7, the scanning square is over a blank square, then either there is a second part to instruction 7, e.g.,

$$7 : * L \; 8 ,$$

or there is no such second part, in which case the computation terminates because the machine cannot go on. The Turing machine owes its fundamental importance to the remarkable theorem that all *concrete* mathematical calculations can be programed on it. ("Concrete" is to be understood in a precise technical sense, but this would involve a discussion of so-called recursive functions and take us a bit too far afield. I here choose to rely on the reader's intuitive feeling for what is meant by concrete.) In other words, every concretely stated computational task will be performed by the machine when it is provided with an appropriate, finite set of instructions.

Because the machine is so primitive, it requires considerable ingenuity to program even the simplest task. For example, multiplication by 2 is accomplished by the following set of instructions:*

```
0 : | * 1
1 : * R 2
2 : | * 3
3 : * R 4
4 : | R 4
4 : * R 5
5 : | R 5
5 : * | 6
6 : | R 6
6 : * | 7
7 : | L 7
7 : * L 8
8 : | L 1
1 : | L 1
```

It should be understood that a number (n) to be multiplied (by 2) is represented by n + 1 consecutive vertical bars, and the computation is begun by placing the scanning square over the bar farthest to the left (instruction 0 starts the calculation).

Figure 2 shows eight consecutive stages of the calculation for n = 3, and the reader should find it instructive to continue to the end to discover the answer to be 3 × 2 = 6.

A real computer is, of course, much more complex and also a much more flexible instrument than the Turing machine. Nevertheless, the art and science of programming a real computer, although sometimes impressively intricate, is very much like that of instructing a Turing machine.

As the reader will have noticed, the instructions must be extraordinarily explicit and complete. The slightest error ("bug") and the result will be hopelessly wrong, the machine will fail to turn itself off, or some other disaster

*Taken from *Theory of Recursive Functions and Effective Computability* by Hartley Rogers, Jr., McGraw-Hill Book Co., 1967.

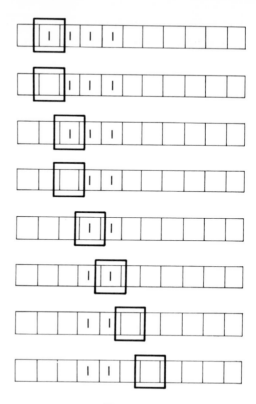

FIGURE 2

will happen. In communicating with the computer there is nothing like the "don't you see?" or "think a little harder" sort of thing. And there is no point whatsoever in getting angry or upset. It will only cause more errors and the computer will go on printing out more gibberish which it "thinks" it was instructed to print out.

Programing calls for a most detailed and rigorous analysis of the task at hand, and "debugging" is puzzle-solving on the highest level.

The difficulties and subtleties of programing are caused to a large extent by the primitiveness of the basic operations that the computer can perform. Could one not improve matters by enlarging the set of fundamental operations? Unfortunately, what might be gained in simplicity and flexibility would be lost in accuracy.

Operations like L, R, or "erase" and "print" are "off-on" operations that require only a flip of a mechanical relay or the opening or closing of an electrical circuit. The only limitations for such operations are cost and speed, and in principle they are infinitely accurate. (This is also the reason for preferring base 2 in computer calculations. To this base, the digits are 0 and 1, ideally suited for "off-on" devices.) Mechanical relays are cheap but also are much too slow, and it was therefore not until they could be replaced by their electronic counterparts that the age of the computer really began.

Actually, the main stumbling block was "memory," i.e., a method of storing information and instructions in a way that would allow for rapid access and recall. This problem was first overcome in the mid-forties with the invention by F. C. Williams of a special memory tube. Improvements and new inventions followed in rapid succession, continuing until this day.

A modern computer is thus a device that can perform extremely simple operations at a fantastic speed and that is endowed with a "memory" in which vast amounts of readily (and rapidly!) available data can be stored for immediate or future use. Instructions are also stored in the memory, but before the computer accepts them ("understands them") they have to be written in one of several special "languages," each designed for a particular use. Some of the most familiar of those languages are FORTRAN, A.P.L., and B.A.S.I.C. This, in essence, is all there is to it. Of course, the detailed way in which everything is put together is intricate and enormously complex.

It would appear that the computer is completely dependent on the man who instructs it. "Garbage in, garbage out" is the contemptuous way in which the computer's total lack of initiative or inventiveness is sometimes described. But this view is not quite just, and the computer can "teach" its masters things they may not have known. To illustrate this I shall recount a story which, although to some extent aprocryphal, is entirely plausible.

Some time ago, attempts were made to instruct the computer to prove theorems in plane geometry. To do this, in addition to giving the computer the axioms and rules of logical inference, one clearly had to provide it with a "strategy." The strategy was simple and consisted of the sole command: "Look for congruent triangles."

Given a theorem to be proved, the computer would first consider all pairs of triangles and, by going back to the assumptions, try to ascertain which of these (if any) were congruent. If a pair of congruent triangles was found, then all conclusions from this discovery would be drawn and compared with what had to be demonstrated. If a match resulted, the computer would proudly print "Q.E.D." and shut itself off. If no conclusion drawn in this way matched the desired one, or if no two triangles were congruent, then the computer was instructed to drop all possible perpendiculars, draw all possible bisectors and medians, and start to search for congruent triangles all over again.

Well, here is what happened when the computer was asked to prove that, given an isoceles triangle with $AC = BC$, angles A and B are also equal [$\angle A = \angle B$]

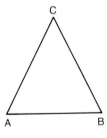

FIGURE 3

The authors of the program fully expected the computer, after a brief contemplation of the lonely triangle ABC, ultimately to drop the perpendicular CD* and, from the congruence of the right triangles ADC and BDC, conclude that $\angle A = \angle B$.

*Students of "New Math." rejoice! The proof requires that D lie between A and B, for which order axioms are needed.

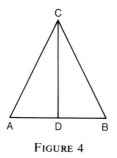

FIGURE 4

But the computer fooled them! It gave a much nicer and a much more sophisticated proof. It simply "noted" that triangles ABC and BAC (!) are congruent and derived the desired conclusion in this way.

Before anyone jumps, let me hasten to state that this proof is not new. In fact, it is very old and was first given by Pappus in the fifth century A.D. It was reproduced in most texts on geometry until about the end of the nineteenth century, when it was decided that it was too difficult for young humans to comprehend.

The reason the proof is difficult is that it goes against the grain to consider ABC and BAC as different triangles when the evidence in front of our eyes makes it clear that we have only one triangle. But the computer cannot see and must therefore rely on definitions and logic. *Formally*, ABC and BAC are *different* triangles, even though they are represented by the same drawing.

Intuition based on sensory experience is sometimes a deterrent to abstract thinking. Therefore, because a computer is deaf and blind, so to speak, it can in some respects and in some instances win over humans.

However, can a computer be truly inventive and creative? This is not an easy question to answer, because we ourselves do not really know what inventiveness and creativity are. One of the more interesting and imaginative lines of research, which goes by the name of "artificial intelligence," is directed toward answering such questions and, somewhat more generally, toward computer simulation of activities usually associated with living beings which may even be endowed with intelligence.

How complex the notion of creativity really is can be illustrated by the following example:

Consider an eight-by-eight-inch board made out of equal (one-by-one inch) squares from which the left uppermost and the right lowermost spaces are removed (Figure 5A).

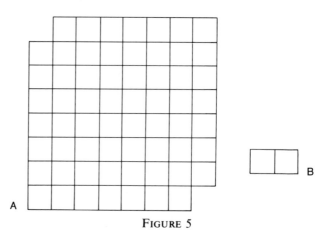

FIGURE 5

Given a set of two-by-one "dominos" (Figure 5B), the question is whether one can completely fill the board by placing them horizontally and vertically in any position without overlapping.

The answer is no, and the proof is as follows:

Imagine that the squares of the board are colored alternately like a chess-board (e.g., white and black) so that any two adjacent squares differ in color. Of the 62 squares, 30 will be of one color and 32 of the other. If we also imagine that one square of the "domino" is black and the other one white, then the set of squares that can be covered by the dominos must, by necessity, have as many squares of one color as of the other. Since our truncated board contains unequal numbers of differently colored squares, it cannot be completely covered by "dominos," q.e.d.

The creativity comes in thinking of coloring the squares with two colors (or, more abstractly, of picking two symbols and assigning differing symbols to alternative squares). There was nothing in the problem as posed to suggest coloring (except perhaps that an eight-by-eight board suggests a chess-board), and the device of alternate colors had to be *invented* for the purpose of proof. Could a machine hit upon such an idea? Even if, by accident, it would hit upon assigning symbols to squares,

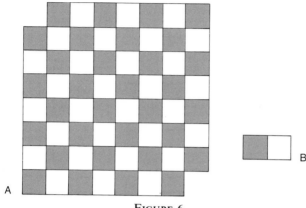

B

FIGURE 6

would it recognize the pertinence of this device to the problem as posed?

The enthusiasts of "artificial intelligence" would unhesitatingly answer in the affirmative. They believe that even a spark of genius is a matter of accumulated experience, and that therefore a machine that has a large enough memory and is capable of extremely rapid collation could, on occasions, pass for a genius. Well, perhaps. But that wouldn't be of much interest if it turned out that, to be creative on the human level, the machine would have to be of the size of our galaxy—or even larger.

As of now, the computer has not even been really successful in performing intellectual tasks that should not be overly taxing. For example, machine translations of relatively simple texts from one language into another have not gone beyond a rather primitive stage, in spite of much effort in this direction. Theorem-proving, despite the (alleged) rediscovery of Pappus's proof, has not progressed far, either, and it may be some time before a computer can compete with a really bright young student.

Still, attempts to simulate feats of human intelligence by machines are interesting and worthwhile, if for no other reason than that they should help us to understand better the intricate workings of our own minds.

However, the list of things the computer can do, and do well, is enormous and it grows longer by the day. Virtually every area of research depends to some degree on a computer. One can now begin to solve realistically approximate equations that describe the motion of the atmosphere. One can simulate in the computer the workings of telephone exchanges, and we should be able to avoid future blackouts affecting large areas by computer studies of the flow of electrical energy in complex networks. Calculation of orbits is routine, and so are many other calculations in physics and astronomy. Space studies are possible because of the vast amount of information poured into computers by satellites, unmanned probes of deep space, and manned space vehicles. Geologists use computers to analyze seismographic tracings that could be forerunners of earthquakes.

In the life sciences, the computer is now an indispensable tool for analyzing X-ray diffraction patterns, which are fundamental in determining structures of such complex molecules as DNA or hemoglobin. It processes huge amounts of neurophysiological data and it helps in testing all sorts of hypotheses and theories in ecology and population genetics.

There are manifold uses for the computer in economics, and in recent years attempts have been made to use it on a grandiose scale to predict the economic future of the world. Although one may question the assumptions underlying the proposed model and remain sceptical about the validity of conclusions drawn from it, the fact remains that an exercise of such magnitude could not have been even contemplated without the computer.

Here, however, a word of caution is necessary. It is precisely because of the power of the computer to simulate models of great complexity and because, in the popular mind, the computer is the ultimate in precision and infallability, that special care must be exercised in warning that the conclusions drawn are merely consequences of the assumptions put in. It is perhaps here that the criticism "garbage in, garbage out" may be justified, unless the assumptions are subjected to a most rigorous and critical analysis.

I could go on and on reciting the uses of computers in engineering, business, and government. There is, in fact, hardly any facet of our lives which has remained untouched by the rapid development of this instrument. But, instead, I should like to conclude by a few remarks about the "intellectual" role of the computer.

It is a difficult, and even a dangerous, task to try to delineate achievements from the instruments that made these achievements possible. Biology and astronomy would not be what they are today if it were not for the microscope and the telescope. And yet neither of these magnificent instruments is what might be called, in an "intellectual" sense, a part of the discipline it serves so indispensably and so well.

It is different with a computer. Although I cannot think of a single major discovery in mathematics or science that can be attributed directly to the computer, in an intellectual sense, the computer as such belongs, without doubt, to mathematics. This is not only because it is rooted in logic, but mainly because it has become a source of new mathematical problems and an instrument of mathematical experimentation. One could even attempt to defend a thesis (although it would be considered ultrasubversive by many, if not most, of my colleagues) that the only mathematical problems which arose during our century and are not traceable to the nineteenth are conceptual problems generated by computers.

The enormously complicated calculations which had to be performed during the wartime development of the atom bomb led John von Neumann, one of the great mathematicians of our time, to design and build one of the first all-purpose electronic computers. His vision, perseverance, and persuasiveness are primarily responsible for the flowering of the computer industry today. But von Neumann was also aware of, and fascinated by, the new world of automata and the novel problems they pose. He predicted, not entirely in jest, that the computer would keep scientists "honest," because logic is a prime requirement in the formulation of a problem that would have meaning for a computer. That is, unless the scientist is himself logical, the computer will be of no help to him at all. And in an intellectual tour de force, he invented a way in which an automaton could reproduce itself (i.e., be capable of reconstructing its own blueprint). Prophetically, this feat bore a more than perfunctory resemblance to the discovery some years later of the way in which living cells accomplish this task.

If computers can inspire such flights of imagination, they can and should be forgiven for some of the silly, irritating, and frustrating things they sometimes do when guided by their imperfect human masters.

CHAPTER 13 V.E. McKELVEY

Building Stones for Affluence, Population Growth, and Environmental Strain

From the beginning of civilization, every nation's basic wealth and progress has stemmed in large measure from its natural resources. This nation has been, and is now, especially fortunate in the blessings we have inherited. Our entire society rests upon—and is dependent upon—our water, our land, our forests, and our minerals. How we use these resources influences our health, security, economy, and well-being.

J. F. Kennedy, Message to Congress on Natural Resources, February 23, 1961.

THE AFFLUENCE of a modern industrial society such as ours stems from the use we make of two kinds of resources: 1) natural resources—the land and its biological products, water, air, mineral materials, and such inanimate sources of energy as oil and coal; and 2) human resources, particularly in the form of the cumulative intellect and ingenuity that has made it possible to transform natural raw materials into products useful to man. The modern farm—in which metals, gasoline, mineral-based fertilizer and insecticides, artificially diverted water, and cleared land have been combined by human ingenuity into a highly productive agricultural unit—symbolizes the intelligent use and transformation of resources that account for our high level of living.

Minerals and fuels are the veritable building stones of the affluent society, because they provide the basis for intensive use of the labor-saving devices and energy that are the unique features of the industrial civilization.

But this very use of natural resources that has led to the material wealth and human comforts of the industrial society also has had two unanticipated consequences of momentous significance. One is that the human population has increased many times over the number that could subsist in a food-gathering society. This is a direct result of the expansion of the carrying power of the land, which was made possible by ingenious use of resources. For example, it is believed that, prior to invasion by Western Europeans, the population of American Indians in the United States was about one million and that it was more or less in equilibrium with the resources available to Indian culture of that time. The population of the United States is now about 210 million and probably will increase by 60 or 70 million by the end of the century. Dr. Segal discussed aspects of this problem in Chapter 6.

The other consequence of our vigorous development of minerals and fuels is that their production and use have contributed to the deterioration of environmental quality. Defacement of the land surface, siltation, and chemical pollution of streams from mining, and air pollution from smelters and refineries are undesirable environmental side effects of the production process. And environmental side effects of the use of minerals and fuels—solid wastes and automotive exhaust, for example—are even more widespread and noxious.

Man's remarkable ability to use natural resources thus has led to a fabulous increase in human wealth and comfort, but it has also imposed a severe strain on the natural environment as a result of both the increase in the number of people that press on that environment and the polluting effects of resource production and use. In man's creative but sometimes clumsy hands, the building stones of affluence have also paved the road to environmental strain.

Use of mineral and fuel resources has increased rapidly in the industrial age and it must continue to increase if still-impoverished people throughout the world are to achieve a high level of living. But are potential resources adequate to support continued expansion in resource production and use? And, if so, can ways be found to reduce the resulting environmental stress in order to maintain a satisfying and viable human environment? In examining these questions, let us begin with a brief review of the building stones themselves—what they are and where they come from.

FROM SUNLIGHT AND FLINT TO ELECTRIC POWER AND EUROPIUM

The mineral sources used directly by primitive man were mainly flint and other rocks from which he could fashion stone tools and weapons, water, and ordinary salt, which was sought as a dietary supplement just as it still is by other animals. His sole source of energy was sunlight, which provided heat and light directly and muscle power indirectly after it was transformed through photosynthesis into edible, energy-yielding foods. Except for the advantages gained from the use of tools and weapons, man's ability to use the earth was comparable to that of other carnivores, and his living was a precarious one indeed.

The subsequent emergence and growth of civilization has been supported—and from a physical standpoint made possible—by the use of a widening spectrum of mineral resources. It is entirely appropriate that historians identify the steps in the advance of civilization by the names of the minerals which made new activities possible. Progress from the Stone Age to the Bronze Age, Iron Age, Coal Age, and Nuclear Age is, in large part, the record of the growth in man's ability to make constructive use of mineral resources.

Until the twentieth century, this growth was slow. In 1900, only about 30 chemical elements were in commercial use. Aluminum was a metal known only in the chemical laboratory. Uranium had just been discovered as a chemical element. Until a decade or so earlier, wood exceeded coal as the principal source of energy. Petroleum furnished only about 2.4 per cent of our energy requirements and the use of natural gas was limited to a few localities near the well head, for pipeline technology had not yet been developed. Only some 5 per cent of our energy was used to generate electricity.

Today approximately 70 chemical elements and 135 mineral commodities are in commercial use in the United States. Aluminum is a common metal. Europium—one of the rare earths—is used in the color television tube, and germanium and other rare metals have made possible the transistor. In 1971, petroleum and natural-gas liquids supplied 44 per cent of our energy, natural gas 33 per cent, coal 18.3 per cent, and hydropower 4.1 per cent. Only 0.6 per cent came from nuclear energy, but nuclear power may be the dominant source of energy in a few decades. About 25 per cent of the energy consumed in the United States is used in producing electricity. Because of the increase in total consumption of energy and the improved efficiency of power generation, the actual amount of electricity generated is now about 275 times what it was at the turn of the century. After all, that was before air-conditioners, radio, television, electric stoves and water heaters, hair driers, electric typewriters, electric refrigerators and freezers, can-openers, and hundreds of other appliances we now take for granted, began to demand electric power.

MINERAL COMFORTS AND CONVENIENCES

Nearly all minerals and mineral fuels are processed, refined, or transformed to other products before use, and the average consumer has little occasion to think about the role they play in supplying the comforts and conveniences of modern life. The automobile and the highway, of course, are made of mineral raw materials, as are the countless machines and motors that are a part of our daily life. The energy to run them comes from mineral

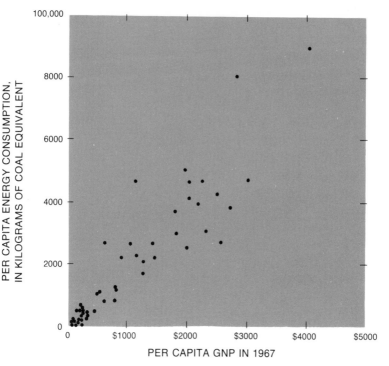

FIGURE 1 *Per capita Gross National Product increases among the countries of the world in rough proportion to their per capita consumption of energy and of minerals.*

fuels. Some indication of the contribution of minerals to affluence may be seen in the fact that a gallon of gasoline burned in a one-horsepower motor yields the energy equivalent of 25 man days of hard physical labor. That is buying human labor at a cost of 1.5 cents a man day!

Looking around at the average American home, one can see a mineral source or component in nearly every article. The concrete, brick, or stone foundation has a stone, sand, gravel, or clay source, and lime is a principal constituent of the mortar. Some of the piping is copper and the rest of it is iron, with zinc on galvanized surfaces. The carpets, drapes, and other textiles are colored mainly by mineral-based dyes. Some of them have a mineral filler, and some are made of synthetic fibers. Synthetic fiber and plastic objects are mainly derived from petro-

211 Building Stones for Growth

leum chemicals, as Dr. Mark pointed out in Chapter 10. The plaster in the walls or plaster board is made largely from gypsum. The paint has mineral pigments and the varnish on the furniture has a mineral drier. The dishes, glassware, silverware, kitchen utensils, and stove are all made from minerals. Even the vegetables in the refrigerator will have benefited from mineral fertilizers and insecticides. All of the electrical appliances, the furnace, and the light bulbs are composed of metals or ceramics and the work they perform is powered primarily by energy from mineral fuels. Most of the medicines are derived from mineral-based chemicals, as are the kitchen chemicals and cleansing agents. Chances are that a close examination would disclose that most of the 70 chemical elements in commercial use today would be found in the home, playing their part in providing the comforts and conveniences that are now an accepted part of American life.

FEAST, THEN FAMINE AND PESTILENCE?

The minerals and mineral fuels that support our high level of living are, for the most part, nonrenewable resources. They are formed by natural processes, but at rates infinitely slower than the speed at which we are consuming them. Once a coal bed has been mined and the fuel burned, the amount of energy available for man's future use is reduced by an equivalent amount. Nonfuel minerals are not destroyed through use, but the effect may be the same if they are used in ways that prevent them from being recycled.

Because of the accelerating rate at which these resources are being used, concern is mounting that they will soon be exhausted. Shortages are developing in supplies of natural gas and gasoline, and domestic production of oil and many other minerals is no longer adequate to supply our needs.

Environmental deterioration also results directly or indirectly from the production and use of minerals and mineral fuels. Many people, therefore, are also con-

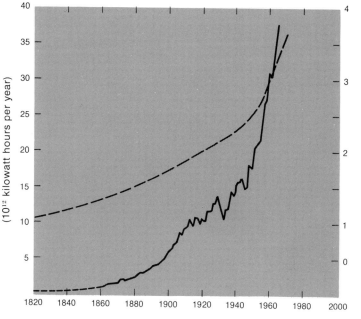

WORLD PRODUCTION OF ENERGY FROM COAL AND PETROLEUM (10^{12} kilowatt hours per year)

WORLD POPULATION (billions)

FIGURE 2 *Increased use of energy (solid line) symbolizes the results of the development of technology that has expanded the carrying power of the land and led to a corresponding increase in population (dashed line).*

cerned that the quality of the environment will deteriorate to the point at which it will no longer support human life. By making the assumption that world resources are equivalent to a 250-year supply at present rate of consumption, Donella Meadows and her colleagues at the Massachusetts Institute of Technology recently concluded in their study of *Limits to Growth* that, at the present exponential rate of increase of consumption, these supplies would be exhausted in less than 100 years. Others have drawn qualitatively similar conclusions. To forestall such a disastrous event, many voices are urging that we reduce our consumption to a much lower level. If we do not do so voluntarily, they say, we may be forced to do so by calamitous exhaustion of supply or environmental degradation.

At exponential rates of growth—such as have been taking place in population, resource consumption, and presumably in environmental deterioration—the quantity involved doubles periodically. At a 7 per cent annual rate of increase—the present rate of increase, for example, in our consumption of electricity—the doubling time is 10 years. At 3 per cent—the rate of increase in population in some countries—the doubling time is about 22 years. Continuation of exponential population growth, no matter at how slow a rate, would inexorably lead to standing room only. And exponential growth in consumption would eventually exhaust any supply, no matter how large. It also is sobering to realize that, in consuming a fixed supply at an exponentially increasing rate, three-fourths of the total will be used in the last two doubling times, no matter how tiny the fraction of the whole was constituted by the first increment of consumption! Warning of impending exhaustion may need to be recognized when only a very small part of any resource has been used.

Obviously, then, exponential growth in population, resource consumption, and environmental deterioration can not continue indefinitely. We must achieve a viable balance of population, resource consumption, and environmental quality, and to do so is a matter of the greatest urgency.

But is it true that resources are so limited and the environmental consequences of their rate of consumption so inevitable that we who have attained a high level of living must immediately seek a much lower one and that the still-disadvantaged peoples of the world can aspire to nothing more?

BRAINPOWER—THE BASIC RESOURCE

We simply do not know enough about our potential resources and the environmental consequences of their use to be able to answer such questions definitively— and it is also a matter of great urgency that, in analyzing problems about our future, we acquire and substitute knowledge for what today are only assumptions. In spite

Figure 3 *This crowded freeway and its smog symbolize many of the accomplishments of the modern world and the problems it has created simultaneously. High use of minerals and energy has resulted in affluence and population growth, accompanied by congestion, air pollution, and deterioration of environmental quality, leading to disruptive and inflationary shortages of minerals and fuels. With little loss in comfort or convenience, we could conserve energy and minerals and reduce environmental impact by using smaller cars, depending more on mass transit, recycling materials, and stabilizing population.*

of information deficiencies, however, we can identify significant potential resources that should carry us much further than those we are using now. We can also identify ways to extend our resources considerably and to avoid or reduce undesirable environmental effects.

As we have already seen from the historical development in the use of minerals and mineral fuels, resources —defined as naturally occurring substances usable by man—are not fixed in quantity, but depend on man's ability to find, mine, refine, transport, and use them economically. When we developed the ability to produce aluminum economically, for instance, our resources of metals were extended greatly, and resources that had not existed in any practical sense literally were created. We can create resources in other ways by advancing our ability to find mineral deposits concealed beneath the earth's surface and to recover, refine, and process materials that previously were inaccessible or unusable. Historically, resources created in this way have been produced at costs no higher, and often somewhat lower, than those that prevailed when needs were satisfied from higher-grade or more accessible deposits that were mined and processed with more primitive technology. Some resources also can be extended by recycling and by increasing the efficiency of their use. We now obtain about 7.8 times as much electric power from a ton of coal as we were able to get at the turn of the century—an accomplishment just as significant as discovering previously unknown coal resources. We can also extend our resources by preventing waste in both production and use. For instance, insulation reduces the energy required for space conditioning, and mass transit is far less wasteful in the use of energy for transportation than is the automobile.

Looking at the problem of future supplies in ways such as these encourages the hope that we can provide the resources to support a good level of living for the world's population far into the future—on the condition, of course, that its growth soon can be controlled. With re-

FIGURE 4 *Petroleum and other mineral resources are ex-tended by scientific and technological advances that permit discovery of deposits in previously inaccessible areas, and that lead to increased recovery and efficiency of use. With care in the planning and conduct of operations, offshore petroleum drilling and production can be undertaken without significant environmental effect.*

spect to energy resources, for example, a substantial potential exists for each of the fuels now in use. Only a small fraction of the rocks that might lead to discoveries of oil and gas have been explored; only about 30 per cent of the oil and 80 per cent of the gas originally in place can be recovered economically now, and petroleum beneath the sea floor in waters deeper than a few hundred meters is not yet accessible. Advances in discovery, recovery, and drilling technology should extend usable petroleum resources considerably. In addition to the potential in deposits of the kind now in wide use are other energy resources that still are little used. Potential resources of hydrocarbons in oil shale and tar sands are far larger than are petroleum resources. Geothermal energy in the form of steam and hot water is just beginning to be explored and developed in the United States. Development of the breeder reactor, described by Dr. Starr in the next chapter, would make available enormous quantities of uranium and thorium in deposits that are too low in quality to be mined economically now. This would increase many-fold the energy that could be recovered from naturally occuring fissionable materials. Control of the fusion reaction for the manufacture of electric power would create even larger resources. And development of the ability to capture solar energy efficiently would give a continuing, inexhaustible supply. Rapid advance in the development of the nonhydrocarbon energy resources would make it possible to save at least some of our petroleum for those nonfuel uses in which it is extremely valuable.

But what about the environmental effects of our high consumption of resources? Can they be avoided? Here also we can identify useful approaches, one of the most important of which is to regard pollution as a form of resource waste. For instance, if the sulfur dioxide from power-plant emissions were captured, air pollution would, of course, be reduced. But in addition, the captured sulfur would be a conserved resource—and its value possibly would help to offset the cost of pollution

FIGURE 5 *This strip mine near Benezette, Pennsylvania, was reclaimed to farmland by backfilling. With careful planning in advance of mining, mined land can be reclaimed efficiently to make it useful for other purposes.*

control. Solid wastes processed to yield the maximum recovery of usable products could prove to be resource assets, instead of environmental liabilities. And strip mining carried out with the intention of restoring the land surface for other use could have a constructive, instead of a destructive, effect on land values.

These approaches to resource and environmental problems identify brain power as the basic resource, the key to man's future as it has been to his past. All of them call for and depend for their success on the advance of our ability to create resources, to extend our knowledge about the earth and natural processes, and to guide and control our own actions to balance our numbers and level of living against resource availability and environmental integrity. Advancement of these abilities constitutes one of the highest challenges of our time. Research has led us to our present high standard of living. More research is needed to help us maintain it. If we respond with vigor and imagination, the answer to the question of whether famine need follow the feast can be determined by man, rather than dictated by nature.

CHAPTER 14 CHAUNCEY STARR

Nuclear Power

> *Every civilization has been grounded on technology:*
> *what makes ours unique is that for the first time we*
> *believe that every man is entitled to all its benefits.*

J. Bronowski, in *The American Scholar*, Spring, 1972

MAN'S MANIPULATION of energy has permitted him to progress from primitive subsistence to a life style that has improved his happiness and longevity. But in the process of so-doing, man has created a considerable amount of waste and harmful by-products, as was pointed out in the previous chapter. These wastes can be managed, and most environmental damages either can be repaired or prevented, but to do so, even more energy will be needed. Technology must provide energy sources that are abundant, relatively inexpensive, and environmentally clean.

The importance of easily available energy can best be described in terms of the per capita needs of the world's population. At the present time, the average citizen of the United States uses the heat-energy equivalent of about 90,000 kilowatt-hours annually for all purposes. It has been projected that future needs for energy—including that for recycling materials, removing waste, and improving the environment—will double this need. The average per capita consumption in the remainder of the world is about one-eighth of that in the United States. Even if the world population were to stabilize at present levels, raising the quality of life in the underdeveloped parts of the world will increase the world-wide demand for energy enormously. If it takes another 50 to 100 years to stabilize the world's population, we may be faced with an even greater increase in demand. Because the world's fossil-fuel reserves are not renewable and are not spread uniformally around the globe, we are already facing serious problems, all of which make more pressing the need for an alternate, long-term energy source. Nuclear power is one means of achieving that objective.

NUCLEAR REACTIONS

There are two types of power-producing nuclear reactions: fusion and fission. The most cosmically important is fusion. It occurs naturally in the sun and stars and involves the very slow creation of heavy atoms from lighter ones, with an associated loss of mass and the resulting production of energy. The second process, fission, depends on the instantaneous breakup of a heavy nucleus into several lighter nuclei, with a net loss of mass. This fission process drives today's nuclear power plants.

In either fusion or fission, energy is produced in accordance with the mass-energy equivalence predicted by Einstein in 1905, familiarly stated as $e = mc^2$, where e is the energy released, m is the mass destroyed, and c is the velocity of light. Energy appears in the high speed of the atomic products and as electromagnetic radiation.

FUSION

Before we go further, perhaps we had better define isotopes—a word that appears in any discussion of nuclear energy. First, the nucleus of an atom is its small core, which is positively charged and acts to hold the atom's orbiting electrons. It can be conceived as made up of neutrons (which have no electric charge) and protons (which are positively charged) in roughly equal numbers. The positive charge of the nucleus determines the chemical properties of the atom. And so we come to isotopes.

A nucleus is an isotope of another if it has the same positive charge but a different mass. Isotopes of one element differ only in the number of neutrons in their nuclei—their chemical properties are the same. Thus, we have as isotopes normal hydrogen with a simple proton nucleus (^1H), deuterium with a neutron added (^2H), and tritium with two neutrons added (^3H)—all the same chemically.

In 1938, Hans Bethe, an American, proposed that thermonuclear fusion was the major source of energy in the sun and stars. He reasoned that at the very high temperature of the sun (20 million degrees C.), atomic nuclei collide with each other at velocities high enough to overcome the electrostatic repulsion of the like positive charges of the nuclei. These high-speed collisions can join four hydrogen nuclei into a helium nucleus. Because 0.7 per cent of the original mass of the hydrogen nuclei is lost, a large amount of energy is released. But the fusion reaction doesn't happen all at once. Even at the sun's high temperatures, the process is slow; only one per cent of the sun's hydrogen is converted in roughly a billion years. If it were faster, the energy production of the sun would not be so constant; as a matter of fact, the sun would have exploded long ago.

Basic research that led to an understanding of the solar fusion processes also led to the conception, about 1950, of the first man-made fusion device: the thermonuclear "hydrogen" bomb, the world's most powerful weapon developed after the A-bomb. Although no explosion could occur if the slow, solar hydrogen cycle were used, several isotopes of hydrogen, particularly deuterium (^2H) and tritium (^3H), do permit rapid interaction. At A-bomb-produced temperatures of tens of millions of degrees, these isotopes can fuse in microseconds.

The successful demonstration of the H-bomb led to an experimental program to develop fusion power for generating electricity. This program is still in the scientific feasibility stage. It may eventually be successful, but is unlikely to be an important power source until at least the year 2,000. The extremely high pressures and temperatures required to sustain a fusion reaction create tremendous materials and engineering problems. However, a substantial international effort is under way to develop controlled fusion reactors.

The first objective of fusion research has been to create in the laboratory conditions similar to those which exist in the sun. The present development has concentrated on devising "magnetic bottles" in which the force of very strong magnetic fields might keep the interacting gases from escaping. More recently, experimentation with a "laser capsule" has been undertaken. In this concept, the interacting gases are completely surrounded by high-energy laser beams which compress and contain the gases for a time sufficient to produce a fusion reaction. Some scientists believe that if this technique can be perfected and controlled, it could produce almost unlimited energy that would be totally free of pollution. If successful, fusion could satisfy a huge portion of the world's demand for energy for thousands of years. (A fuller discussion of the laser is given in Dr. Schawlow's Chapter 11.)

FISSION

Of more immediate concern is nuclear fission. Based upon the interactions of neutrons with heavy atoms, such as in uranium, fission is the only nuclear reaction now available to produce electricity. This man-made process is the result of a combination of exciting and independent scientific discoveries that were not initially related to any purpose other than an attempt to understand nature.

The scientific structure behind nuclear fission is founded upon the discovery of the neutron by the British physicist James Chadwick in 1932. As we have noted, a neutron is a particle with the same mass as a proton and, most importantly, is without a net electric charge. For this reason, the neutron can penetrate the nucleus of an atom, whose barrier of electrostatic repulsion tends to keep protons and other atomic nuclei out. This penetration is possible even at low velocities, and the process is called "neutron capture." When a neutron is captured by an atomic nucleus, the resulting combination is usually unstable, with consequent production of radioactive emissions of various types until a stable configuration is restored. This phenomenon of neutron-induced radioactivity was first dicovered in 1934 by the French team of Frédéric and Irène Joliot-Curie, and opened up a new field of scientific study on the properties of the nucleus and of the artificially created radioactive elements.

One of the scientists fascinated by this idea was the brilliant Italian Enrico Fermi. In that same year, Fermi bombarded uranium with neutrons. He found that the radioactive elements created by the bombardment did not follow the pattern of natural radioactive decay (the gradual sequence of emissions from an unstable nucleus until a stable one is reached). Fermi assumed that he had produced elements heavier than uranium by adding neutrons to the uranium nucleus, and that their radioactive decay was the source of this unusual mix.

In 1938, Otto Hahn and Fritz Strassman, in Germany, repeated Fermi's experiment and chemically separated the atoms that were produced. They found to their surprise that the uranium atom apparently had been split instantaneously into new atomic nuclei. In Sweden, Lise Meitner, who had at one time collaborated with Hahn, heard of Hahn's latest experiment and, with Otto Frisch, correctly interpreted the new results. What had been observed was "fission"—the splitting of the uranium nucleus into two parts. That possibility had never before been considered seriously. Their findings were communicated to Niels Bohr at the Institute of Theoretical Physics in Copenhagen, and he revealed the remarkable nature of this new discovery.

It was recognized that the disruptive process in uranium would be associated with a large energy release because of mass differences. Thus, early in 1939, the world was informed of a fundamentally new nuclear reaction and energy phenomenon. The scientific excitement was so great that almost 100 papers were published on nuclear fission in the following year. The advent of World War II muzzled open discussion of the topic.

Although it was of great scientific importance, and a truly fundamental discovery, the fission process as an energy source became possible only when it was discovered that, in addition to the two atomic particles (the so-called "fission products") resulting from the splitting of the uranium nucleus, fission also produced two or more neutrons of very high energy (Figure 1). Scientists were excited about the possibility that the fission process, once begun, might be able to sustain itself. It was clear that if one or more of the new neutrons could be captured by other uranium nuclei, the number of atoms fissioning would multiply and thus produce a chain reaction that would continously multiply the energy released (Figure 2).

The fission energy primarily exists in the velocity of the fission products, and secondarily in the velocity of the neutrons and emitted gamma rays and electrons. This energy is converted to heat when these particles collide with surrounding materials—a kind of friction similar to that which causes automobile brakes to heat. The fission of the atoms in a single gram of uranium 235 releases

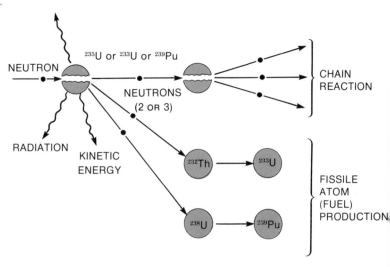

FIGURE 1 *Chain reaction.*

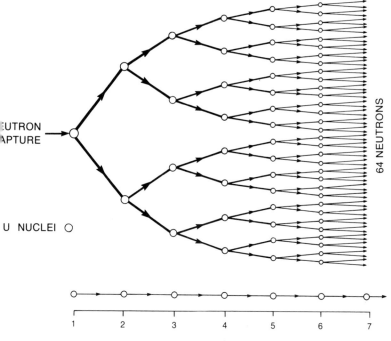

NEUTRON
CAPTURE →

U NUCLEI ○

64 NEUTRONS

1 2 3 4 5 6 7

FIGURE 2 *Neutron chain reaction, or neutron generation. In a two-for-one reproduction ratio (top) neutrons and fissions multiply quickly, as two neutrons become available for each one captured. In one-for-one ratio (open circles and arrows, bottom), one neutron is available for each one captured. In that way a chain reaction is kept going at a steady level to produce constant energy.*

about 23,000 kilowatt hours of thermal energy, equivalent to that from burning about $2\frac{1}{2}$ million times as much coal or oil.

One of the key questions was which of the two known isotopes of uranium, 238 or 235, was responsible for the splitting. These two isotopes had been discovered in 1935 by Arthur Dempster, an American. Uranium 235 is rare; it occurs in natural uranium as only 0.7 per cent of the total. Unfortunately, this rare isotope is the one that captures neutrons and fissions most readily. The uranium 238 atom also captures a small fraction of the neutrons, but generally it does not split (Figure 2). This means that

in natural uranium the excess neutrons produced by fissioning ^{235}U atoms are captured by the ^{238}U, thus killing the chain reaction. Fortunately, uranium 235 captures neutrons easily if they are moving very slowly past the uranium nucleus, whereas the ^{238}U nucleus does not. The new neutrons can be slowed by "billiard-ball" collisions with nonabsorbing material, such as the hydrogen nuclei in water or the carbon nuclei in graphite. Such "moderation" of the neutron velocities increases their probability of capture by ^{235}U and thus creates more fissioning. The neutrons that collide back and forth with the moderating material at room temperatures are called thermal, or slow, neutrons (Figure 3).

NUCLEAR REACTORS

After these concepts were developed in 1939, the experimental and theoretical studies for the use of nuclear energy took two directions, both as military programs under the "Manhattan Project." Secret laboratories at the universities of Chicago, Columbia, and California and at the new sites of Los Alamos, New Mexico; Oak Ridge, Tennessee; and Hanford, Washington, became beehives of activity. The greatest scientists of the United States, Canada, and England were assembled to guide the work. One program involved the building of a large "atomic pile" of natural uranium rods spaced between layers of graphite "moderator." The chain reaction that resulted when atoms of ^{235}U were fissioned by the slowed neutrons was controlled by introducing neutron-absorbers, which regulated the number of neutrons circulating in this "pile." The term "atomic pile" has now been superseded by the term "nuclear reactor" (Figure 4).

The second direction was the attempt to develop an atomic bomb. It was evident that, for an explosive reaction, the neutrons would have to multiply very rapidly, and that this could happen only if relatively pure ^{235}U could be used. Thus a massive program was developed to separate ^{235}U from ^{238}U.

In December of 1942, under the leadership of Enrico

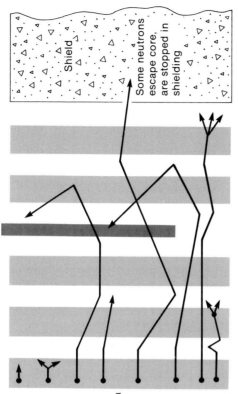

Fuel element
Moderator

Control rod

Shield

Some neutrons escape core, are stopped in shielding

Of the fast neutrons born in fission in a fuel element...

...A few cause additional fissions before they slow down...

Most are slowed down by collisions with atoms of the moderator...

Some of these are absorbed while slowing down...

The rest come to thermal equilibrium with and diffuse through the reactor core...

Some of these diffusing neutrons are captured in moderator or core structure...

The rest, enough to keep the chain reaction going, are captured in fuel and cause fission, producing the next generation of neutrons...

FIGURE 3 *The controlled chain reaction—a typical neutron "generation." The reactor designer's problem is to insure that, on the average, one of the approximately 2.5 neutrons released by the fission of a uranium nucleus survives to cause fission in yet another ^{235}U nucleus. The essential challenge is to do this economically while meeting structural and heat-removal requirements.*

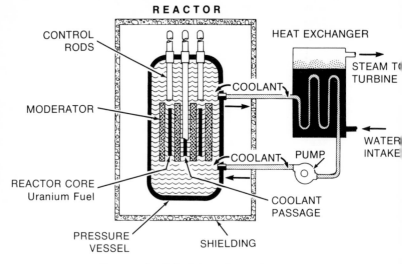

FIGURE 4 *Principles of a nuclear reactor.*

Fermi, the first sustained nuclear-fission chain reaction took place. This historic occasion verified many of the scientific and intuitive estimates that had been made concerning the nature of the fission process, the number of neutrons released, and their capture in materials.

It also opened up the way for the massive production of plutonium (Pu) 239, an easily fissionable element produced fractionally when ^{238}U captures neutrons. With the development of a workable nuclear reactor and the resulting ability to make a huge number of neutrons, the possibility of producing plutonium from ^{238}U became feasible. The immediate outcome of the research was the A-bomb, initially based on ^{235}U and, subsequently, on ^{239}Pu. In addition, the military program provided the technological base from which scientists developed today's nuclear-powered generating plants.

The basic problem in all nuclear-reactor design is the balance between the production of neutrons in the fission process and the absorption, or waste, of neutrons in various other materials in the reactor core, including the natural competition between ^{235}U and ^{238}U. This can be adjusted by increasing the relative amount of fissile ^{235}U (or ^{239}Pu) in the reactor core. Thus, a reactor that contains fissionable isotopes in excess of the natural ratio is called an "enriched reactor."

Maximum fuel enrichment and complete removal of the moderator results in a chain reaction based on fast neutrons and is a design called a "fast breeder." In the fast breeder, the number of neutrons in excess of those needed for the chain reaction is so great that those neutrons leaking the reacting core can be absorbed usefully in a surrounding blanket of ^{238}U to breed ^{239}Pu, or in a blanket of thorium to breed ^{233}U (Figure 1). ^{233}U is also easily fissionable. When the fissile fuel that is bred is more than the amount used to produce power, a "breeding cycle" has begun. Experimental fast breeders are now in operation in several countries.

The nuclear power plants that produce electricity in the United States are supported by a complex network of processing facilities (Figure 5). This complex can only be

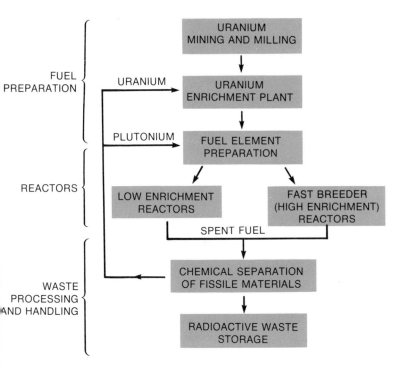

FIGURE 5   *Uranium-plutonium fuel cycle.*

justified if nuclear power is important to the world's social development. This is particularly important because a by-product of nuclear power is the production of large amounts of radioactive wastes. While the volume and mass of such radioactive wastes are small compared to the power produced, they are cancer-producing, even in microscopic quantities, and therefore must be handled and processed with extreme care.

NEED FOR NUCLEAR POWER

Is all of this expense, research, and development worth the price and effort? To answer that question, let us examine nuclear power as an energy source for society. Without question, the world is running short of easily available fossil fuels (coal, oil, and gas). Not only is their availability dropping; their cost is going up. On the other hand, energy from uranium ore is certainly as great or greater than that from fossil fuels. More dramatic is that the same uranium resource, when utilized in the breeding cycle, will permit a nearly 100-fold increase in the availability of low-cost energy.

Yet there continue to be many technological problems in developing reliable and economical nuclear power plants. The issue of public safety clearly deserves special consideration. There are three sources of public exposure to excessive radiation from the operation of nuclear power plants. The first is the routine operation of the plant under its normal condition of design. The second is the possibility of excessive radioactive emissions as a result of accident. The third can arise from the handling and storing of radioactive fuel and waste. Let us take each of these in turn.

The steady-state emissions from nuclear power stations are now extremely low as compared to the normal "background radiation" to which man is constantly exposed. Radioactive minerals and radiation from outer space have always been part of man's natural environment and, over the centuries, may have caused a number of natural mutations. However, the routine emission now permitted from nuclear power stations represents less than one per cent of these natural radiations.

We do know that radiation a million times higher than permissible routine levels can be lethal. Therefore, the exposure which might result from extreme nuclear accidents has been a serious subject of study. Our navy has used nuclear power to operate various vessels for an equivalent of about 1,000 reactor years; about one-tenth that much has been used commercially. During this period of about twenty years, there has not been a single accident in a nuclear power plant that has released radiation to the public. Present studies on the probability of a large-scale accident indicate that it might happen once in 10-million reactor years of operation. The reason lies in the complex, multiple sequence of backup safety devices in nuclear power stations, each of which takes over if the preceding one has failed.

Unquestionably, any risk to human health and life is a serious matter. Yet, we must also recognize that the alternate sources of power, especially the fossil fuels, are themselves creators of pollution and are, in fact, responsible for some of the respiratory diseases we find in urban populations. Also, accidental fires are associated with the large oil storage and pumping areas that supply present power plants. A statistical study of the past effects of effluents from power plants that use fossil fuel and of the past fires in oil-storage areas indicates that these power stations may be considerably more dangerous to the public than those powered by nuclear energy. Of course, one weakness of such a comparison is that fossil-fuel data is written in history, whereas statistics on accidents from nuclear stations remain hypothetical. The important point is the clear and present disadvantage of continuing to use fossil-fuel power stations as our principal energy source. There is also a clear and present advantage in using nuclear power because of its comparatively low environmental impact and its "clean air" qualities.

FIGURE 6 *Two domes, one of destruction, one of construc-*
tion. At the left, empty warships are dwarfed by cloud from a
nuclear test at Bikini atoll, July 24, 1946. At right, the San

RADIOACTIVE WASTES

Public discussions of nuclear power and public safety oc-
casionally focus on the dangers of storing radioactive
wastes for long periods of time—even for centuries. Undoubtedly, such long-term storage is an essential part of
nuclear-power development, but the projected public-
safety issue involved is minimal, compared with other
environmental problems. The fact is that a completely
adequate waste-storage system is trivial in scope and
cost, although not in importance.

The average citizen of the United States now con-
sumes about 7,000 kilowatt hours of electricity per year,
which, it so happens, is the same amount of power obtained from the fissioning of about one gram of nuclear
fuel, equal to the weight of three aspirin tablets but
smaller than a single aspirin. The safest way to handle
the waste mixture is to drip it in liquid form into a small
ceramic pot, heated by electric coils, where it boils dry

Onofre Nuclear Generating Station at San Clemente, California, provides 430,000 watts of energy—enough power to supply a city of more than half-a-million population.

and then melts to form a glass-like ceramic clinker, which is insoluble in water. This process is carried out in a sealed chamber behind heavy walls, watched with telescopes or periscopes, and operated by remote controls. After firing and cooling are completed, the clinker, pot and all, is sealed up inside a can, which, in turn, is sealed inside another. It then can be moved safely in a thick-walled shipping cask to a final storage place. (The casks are designed to survive all possible expected shipping accidents.) In clinker form, the one gram of fission products and its accompanying inert material (representing about one man-year of electricity use) will occupy approximately 1/10 of an ounce, volume measure, and the total cost of processing, transporting, and typically permanent storing will amount to roughly 14 cents. The radioactive energy emitted is 1/40 of a watt, or about 1/10 the energy of a pen-light flashlight.

But the question remains, where *would* we store such

wastes? How can we contemplate a continuity of protection and integrity of containment that will extend over hundreds of generations? Has anyone ever made such commitments before? The answer is yes—in Egypt, and with rather remarkable success. Wooden chests and sarcophagi removed from the Egyptian pyramids are perfectly preserved and look like new wood after 5,000 years in the desert. The Egyptian metal, ceramic, and glass objects are, of course, also unchanged. Can we not do as well?

The great pyramid of Cheops is about 750 feet square at the base. A similar structure could be arranged to form a series of smaller vaults that would house the wastes of the present generating capacity of the United States, if entirely nuclear powered, for more than the 5,000 years the pyramid protected its relics. By then, some spent waste could be removed to make room for new, so that, in fact, a perpetual capacity would exist for our present rate of electricity use. New "pyramids" would be needed as loads increased, perhaps one every decade or so. To compare with such a 750-foot-square pyramid, consider the equivalent energy in strip-mined coal, which would mean handling a volume of fuel and waste 750 feet in width, 300 feet deep, and 200,000 miles long!

In summary, nuclear power is a development growing out of a history of pure scientific research into the basic structure of the nucleus. It may represent one of the greatest developments of man—comparable to the use of fire—by providing a completely new and important energy resource. At the very least, nuclear power, even with present commercial nuclear reactors, could double the world's available depletable resource. At the very best, the fast-breeder cycle could permit this resource of fissionable fuel to be used for the next thousand years. One would hope that, at some future time, the development of the fusion reactor or the efficient utilization of solar radiation will provide mankind with even greater resources for meeting the world's energy needs.

CHAPTER 15 JOHN R. PIERCE

Global Communication

> It is not the present moment which is of interest to the
> scientist or to those who look over his shoulder . . . . It
> is the future.
>
> James B. Conant, in *Science and Common Sense*

WHEN MARCO POLO traveled from Italy to China in the
thirteenth century, the journey took three years. At first,
no one believed his accounts of the wonders he had seen
during his trip and in the court of Kublai Khan, and his
journal was not published until 600 years later.

Today, millions of us take for granted our remarkable
ability to see people and events in all parts of the world
instantly merely by switching on our television sets.
Unlike Marco Polo's unbelieving listeners, we cannot
doubt what we see—even though we often don't know
what to make of it.

The speed of travel and communication increased very
slowly after Marco Polo's time. In the eighteenth and
early nineteenth centuries, large sailing ships touched
ports around the world and, later, steamships answered
the demand for greater speed. Yet, it was only after electrical communication began with the telegraph in the
middle of the last century that it became possible to communicate globally with the speed of light. And the telegraph had to wait for the discovery of new laws of nature.

The fundamental discovery that led to the telegraph
was made in 1819 by Hans Christian Oersted, a Danish
scientist who found that an electric current sets up a

magnetic field which causes a magnetic needle to react. That is, an electrical action produces a mechanical reaction. Building on this finding, Samuel F. B. Morse, an American, developed the first telegraph that was successful commercially. On May 24, 1844, the first telegraph message was sent from the United States Supreme Court to Baltimore—the famous "What hath God wrought!" In 1858, the year of the Lincoln-Douglas debates, the first transatlantic cable was laid (Figure 1). It had been developed by Cyrus Field, another American, and in August of that year, Queen Victoria of England sent a 98-word message of congratulation to President Buchanan—a message that took $16\frac{1}{2}$ hours to transmit. The President's 149-word reply took 10 hours

FIGURE 1 *In late May, 1858, the second telegraph fleet assembled in the Bay of Biscay, an Atlantic Ocean inlet on the west coast of France and the north coast of Spain. There they rehearsed cable splicing and testing the improved machinery that had been newly designed to avoid line breaks. In 1865, another try at the cable was made, and was successful.*

to send. Excitement ran high throughout the world. One enthusiast declared: "Columbus said, 'There is one world, let there be two'; but Field said, 'There are two worlds, let there be one.' " The cable failed a month later, but Field tried again in 1865, and by 1866 it was in operation.

The success of the cable rested in large part on the careful analyses, inventions, and measurements of the English physicist William Thomson, who later became Lord Kelvin. Thomson showed mathematically how a short electrical pulse applied to one end of a cable lengthens as it travels to the other end, thus limiting signaling speed. He invented two highly sensitive receivers, the mirror galvanometer and the siphon recorder. He also measured the resistance of the copper used in Field's early cables and found that a better and more uniform quality was needed. By the beginning of World War I, in 1914, the earth was circled by submarine telegraph cables that carried business and diplomatic messages to most peoples of the world.

Today telegraphy seems a slow and impersonal means of communication, but it was less than 100 years ago, in 1876, that Alexander Graham Bell, teacher of the deaf and student of speech, invented the telephone and voice communication became a reality. It took many years for the telephone to span the ocean as the telegraph had. It became possible only as the results of research conducted by a number of scientists in a number of different countries were interwoven to produce radio telephony.

In 1864, James Clerk Maxwell, a Scottish physicist, had published his theory of electricity and magnetism, which predicted the existence of electromagnetic waves. In 1866, Heinrich Hertz, a German physicist, showed that such waves could be generated, travel through space, and be detected. Eight to ten years later (1894-96), Guglielmo Marconi, an Italian electrical engineer and inventor, demonstrated *wireless* telegraphy by means of electromagnetic (radio) waves.

The radio telephone, which combined the previous

findings, required something more, and that, too, was supplied by science and invention. In the 1880s, a number of experimenters, including Thomas A. Edison, observed that currents in a vacuum pass in one direction from a hot filament to a cold plate. In 1897, Sir J. J. Thomson, a British physicist, showed that such currents are carried by particles called electrons, and he measured the ratio of their charge to their mass.

Early in the twentieth century, J. A. Fleming, also British, made a detector for radio waves, and about 1906, Lee De Forest, an American inventor and radio pioneer, added a screen, or grid, between the hot filament and cold plate. A small voltage on the grid could control the current flowing through the tube. This triode tube could generate continuous radio waves and could amplify and detect them, thus elaborating on Hertz's demonstration.

Radio telephony spanned the Atlantic in 1915, began as a regular service between England and the United States in 1927, and quickly spread over the earth. Radio telephony between continents was, of course, an important advance in global communications, but the quality was often poor, so that conversations were wearying and exasperating. The first high-quality transoceanic telephone service came with the laying of a telephone cable between America and the British Isles in 1956 (Figure 2).

The submarine telephone cable, which lies at the bottom of the ocean, is the outgrowth of long years of work by many types of scientists and engineers. Such a cable can carry telephone messages only because the signals are amplified by repeaters. These repeaters use power that is sent over the cable itself. In the cable laid in 1956, the repeaters used electron tubes based on De Forest's invention and spaced 37.5 nautical miles apart. That cable is still in operation, but it can carry only 48 telephone channels.

The transistor is a more efficient amplifier than is the vacuum tube, and is an excellent example of the way in which scientific findings from many lands can produce

FIGURE 2 *In 1955, the first cable to deliver high-quality telephone service was laid between North America and Oban, Scotland. Here it is being pulled ashore at Clarenville, Newfoundland. It was floated on shallow steel drums, loaded into a trench, and routed to the shore station. The cable was 1,940.8 nautical miles long and began operation at 11 A.M., EST, on September 25, 1956.*

technological innovation. The transistor was made possible by an understanding of quantum mechanics, which was developed by such noted physicists as Albert Einstein and Max Planck, who were Germans, the Austrian Erwin Schroedinger, and the Dane Niels Bohr, all of whom became Nobel Laureates. The work of these men led to an understanding of how electrons and atoms really behave—knowledge that enabled the British physicist H. H. Wilson to explain in 1934 how electrical charges can travel through semiconducting crystals, such as silicon and germanium.

In 1948, three Americans—William Shockley, John Bardeen, and Walter Brattain—discovered that negative charges (electrons) and positive charges ("holes," or absences of electrons) can exist simultaneously in a semiconductor. A thin layer in a semiconductor, a layer made by conducting through holes, acts as a "grid" through which electrons pass, and the grid can control the flow of electrons. This is how the transistor operates (Figure 3). In 1956, Shockley, Bardeen, and Brattain received a Nobel prize for their work.

A submarine cable that was laid that same year makes use of transistors. Because these require so little power, the repeaters can be spaced only 10 nautical miles apart and the system can carry 800 telephone channels. A system to be put in service in 1976 will have repeaters every 5.2 nautical miles and will carry 4,000 telephone channels. Today more than 20 submarine telephone cables are in operation. They form a network around the world and carry messages of all sorts, from voice communication to data fed directly into a computer.

Undeniably, submarine cables are useful, but it was radio communication via satellite that first gave us transoceanic television. This was long in coming for the simple reason that the earth is spherical. Radio waves tend to travel in straight lines, so we might conclude that it would be impossible to use radio waves to communicate between distant points. Such waves would not follow the curve of the earth, but instead would simply shoot straight out into space. However, the earth is surrounded by a high, ionized layer called the ionosphere. The presence of this layer was deduced by Oliver Heaviside, an English electrical engineer and mathematician, and a Harvard professor, A. E. Kennelly. (The ionosphere was first named the Heaviside layer and then the Heaviside-Kennelly layer, before its present name came into general use.) The ionosphere reflects radio waves that have frequencies below about 30 million

FIGURE 3   *This is the first transistor ever assembled; the year was 1947. It was called a "point contact" transistor because amplification or "transistor action" took place when two pointed metal contacts were pressed onto the surface of the semiconductor, which is made of germanium and rests on a metal base. The contacts, made of gold, are supported by a wedge-shaped piece of insulating material. They are placed extremely close together so that they are separated by only a few thousandths of an inch.*

Hertz—the name given to cycles per second, in honor of Heinrich Hertz. Thus, radio waves of low enough frequency bounce back and forth between the ionosphere and the earth and so make their way around the world.

Such "short-wave" radio often gives poor-quality transmission, and it cannot transmit television signals because visual signals consist of a wider range, or *band*, of frequencies, than do speech signals. This makes it necessary to use higher-frequency radio waves for their transmission. However, microwaves—radio waves whose frequencies lie above three billion Hertz—are suitable for the purpose.

The exploitation of microwaves was made possible by many advances, including new forms of De Forest's triode, the invention of the klystron by Russell H. Varian shortly before World War II, and the invention of the traveling wave tube by Rudolf Kompfner later in World War II. The latter tubes can amplify signals at far higher frequencies than can triodes and would one day be used in all communication satellites. A microwave system that transmitted television spanned the United States by 1951. Signals were sent from one hilltop tower to another and the towers were spaced about 30 miles apart, so the top of each tower was visible from the top of the next. This worked for continental United States, but how could microwaves be sent across the ocean?

Because the atmosphere is uneven and sometimes has layers or small regions of different density, it was known that a very little of the power of a radio wave aimed above the horizon is scattered down to a distant point. Such tropospheric scatter, as it is called, could provide some communication up to distances of a couple of hundred miles, but it is not good enough for television transmission.

The space age was approaching, and in 1958 Walter Morrow and his associates in Lincoln Laboratory at the Massachusetts Institute of Technology devised an ingenious scheme for scatter communication. This was called Project West Ford. Their proposal was to launch in orbit around the earth a belt of short wires, which would scatter microwave signals back to earth. Astronomers, optical and radio alike, immediately objected to the cluttering of space with large volumes of material that might interfere with observations. These early environmentalists protested the pollution of space. To meet their protestations, wires were devised that would disintegrate with time. In 1963, 20 kilograms (about 44 pounds) of wires were indeed launched. Then Project West Ford was abandoned. This was less because of marginal results or objections by astronomers and environmentalists then because something better had come along.

That something better was communication by means of earth satellites, which are simultaneously visible from widely separated points on the earth's surface. Microwaves, which penetrate the ionosphere, can be sent in a straight line to such a satellite and thence in another straight line to a point far distant on the earth. In a sense, a satellite is an enormously tall relay point in the sky.

Of course, the moon is a natural satellite of the earth, and Apollo missions demonstrated the ease with which TV signals can be sent here from the moon. This was known theoretically in 1943, but then there was no way to get radio equipment to the lunar surface, some quarter of a million miles away. One could, however, use the moon as a passive reflector, rather like a shiny Christmas-tree ornament. A small part of a radio signal beamed at the moon would be reflected back to earth. As early as 1954, the U.S. Naval Research Laboratory transmitted voice signals by reflection from the moon, and in 1959 an operational moon-link was established between the United States and Hawaii. But the reflected signal is weak and the moon is not always above the horizon. Worse than this, the moon is rough with high mountains and deep craters, so radio waves reflected from different parts of the near side of the moon come back at different times. As a result, the moon provides only a little communication, and that chiefly of teletypewriter and other data signals.

In 1945, Arthur C. Clarke, physicist and science-fiction writer, published an article in which he predicted that some day we would have continuously manned space stations in orbit around the earth (we still don't). He proposed that such space stations could be put in orbits 22,000 miles above the earth's equator, where they would hang above one point on the earth's surface as they revolved and the earth rotated once each 24 hours.

Clarke proposed that such manned space stations be equipped with radio receiving and transmitting equipment so they could broadcast directly to or relay messages between various distant points on earth. Such a manned communication satellite has a great deal in common with the unmanned synchronous communication satellites that now hang in one spot and carry voice and TV signals all over the world, but many steps had to be taken before the present satellites could be launched.

In 1955, when I was with the Bell Telephone Laboratories, I published a paper proposing *unmanned* communication satellites. The paper discussed both passive satellites, such as metalized balloons, and active satellites, containing radio receivers and transmitters. It also proposed satellites revolving in low orbits and others in high synchronous orbits.

The National Aeronautics and Space Administration (NASA) was developing rapidly. They seized on the satellite proposal and on August 12, 1960, launched a 100-foot, shiny, metalized, plastic balloon called Echo (Figure 4A and B). During its first circling, Echo reflected a recorded message from President Eisenhower across the continent from about 1,000 miles above the earth. Echo was used in experimental facsimile and data transmission, and signals reflected from the balloon were picked up in Europe. It was visible on all continents as a swiftly moving, brilliant star in the sky, and public interest in it was so great that a number of newspapers listed the times when it would be visible locally. Echo's chief value, aside from exciting interest, was that it showed that satellite communication was indeed feasible.

Spurred on by Echo, the American Telephone and Telegraph Company designed the Telstar satellite and paid NASA to launch it. Telstar was sent into orbit on July 10, 1962 (Figure 5). Like Echo, the new satellite was launched in a low, nonsynchronous orbit, chiefly because of the limitations of available launching vehicles, and so was visible in both America and Europe for less than half an hour at a time. But Telstar was an active satellite,

FIGURE 4A *On June 5, 1783, in France, man left the earth for the first time when Joseph and Etienne Montgolfier rose to 6,000 feet in their balloon, which was 36 feet in diameter.*

FIGURE 4B *Project Echo satellite was a plastic sphere, 100 feet in diameter. It was launched in August, 1960, 177 years after the Montgolfiers' feat. When in orbit some 1,000 miles above the earth, the thin plastic with its aluminized surface reflects microwaves.*

powered by the sunlight falling on solar cells, and it carried both a radio receiver and a radio transmitter. It could and did relay television pictures as well as voice signals between the United States and Europe. The TV programs created a sensation, and satellite communication was accepted as a practical reality.

However, the problem of launching a satellite in synchronous orbit still remained. In 1960, the Advanced Research Projects Agency (a division of the Department of Defense) set in motion Project Advent, which was to have launched a huge satellite into synchronous orbit. The project cost $170 million, but two years later it was abandoned because the goals of Advent outran the technology that was then available.

The problem was solved by Harold A. Rosen of Hughes Aircraft. He devised an extremely simple, lightweight satellite that could be sent into a synchronous orbit by using the same Thor vehicle that had launched Echo and Telstar. He interested NASA in supporting his ideas, and the first successful synchronous satellite, Syncom II, was launched on July 26, 1963 (the launch of Syncom I had failed).

After Echo was launched, animated discussion began concerning who should be allowed to launch and use satellites. In 1962, the Congress passed the Communication Satellite Act, which vested the sole American right to launch satellites for international communication in a new corporation—the Communication Satellite Corporation, now known generally as Comsat. Although it is a private corporation, some of Comsat's directors are appointed by the government.

Comsat tried to deal individually with various foreign communication agencies, but these banded together into an association called International Telecommunications Satellite Consortium—Intelsat, for short—in which Comsat represents the United States. Intelsat now includes some 70 countries and has launched increasingly larger satellites, which hover more than 22,000 miles above the earth (the distance Arthur Clarke had proposed for manned space stations almost 30 years earlier).

Intelsat I, known familiarly as "Early Bird," was much like Syncom. It was built by Hughes Aircraft and launched in 1965. It provided 240 two-way voice telephone circuits or two television channels. The Intelsat II satellites were slightly different but had essentially the same capacity. The Intelsat III satellites, built by Thompson Ramo Wooldridge Corporation, are more sophisticated, and can handle 1,200 two-way voice circuits or two TV channels. Next came the Intelsat IV satellites, built by Hughes, which provide 12 one-way TV channels or 5,000 two-way voice circuits.

We now have data, telephone, and television communication to almost all parts of the world via satellite (Figure 6). Indeed, present plans call for the NASA Advanced Technology Satellite F to be used experimentally to broadcast television programs directly to villages in India—villages not connected to one another by ground-based communication circuits.

And global communications continue to grow rapidly. Overseas telephone calls increase at a rate of some 30 per cent a year—a factor of 10 in less than 10 years. On our TV screens we have watched, as they took place, children starving in Biafra, the invasion of Czechoslovakia, the war in Vietnam, student demonstrations in France and Japan, strikes in England, revolt in Northern Ireland, and the President's trips to mainland China and to the Soviet Union. Is this good or bad? What is it doing to us and our world?

A strong case can be made that instant, world-wide communication can have bad effects as well as good. It can distract our attention from local problems, which we could do something about, to distant problems we, as individuals, cannot solve. It can inflame people about remote conflicts that might otherwise pass unnoticed. Instant global communication may even be an important factor in causing and maintaining wars in far lands.

One common theme runs through the strange, foreign

FIGURE 5 *Telstar, launched on July 10, 1962, was the first satellite to carry TV programs across the ocean. It was 32 inches in diameter and weighed 172 pounds. It was built by the Bell Telephone Laboratories and launched by the National Aeronautics and Space Administration. This is a model of Telstar II, against an artificial background.*

world we experience through TV because of the technology of instant global communication. People in distant countries live very different lives and believe very different things. Yet whatever their philosophies or life styles, they want our successful science and technology to become a part of their culture. They want the things they have learned about, often through television. They want automobiles, electric power, decent living quarters, and cheap, machine-made consumer goods. The wisest among them also want better agricultural methods, better varieties of food plants, better health care, and better methods of education that science can or should provide.

Yet, many Americans, particularly in recent years, have expressed skepticism about the value of science. Some feel that science and technology are tainted if they arose through research done for military purposes. Surely, instant, world-wide communication is here to stay—and most of us would not care to do without it. We cannot reject it simply because the first effective rockets were developed for military purposes in Germany or because the vehicles used to launch present communication satellites started out as ballistic missiles. If we can solve our pressing educational problems, it may well be through the early efforts of psychologists who worked on developing new procedures for military and industrial training; who discovered methods of teaching that are more effective than those used in many elementary schools today.

Science and technology have diverse origins, but nonetheless represent a continuum of progress. The vehicles that launch communication satellites would have been impossible without the fundamental knowledge of motion and mathematics that Isaac Newton gave us in the seventeenth century. They would have been impossible without a host of mechanical, aerodynamic, chemical, and electrical discoveries made in universities, research institutes, and government and industrial laboratories since Newton's time. Communication satellites themselves would have been impossible without the

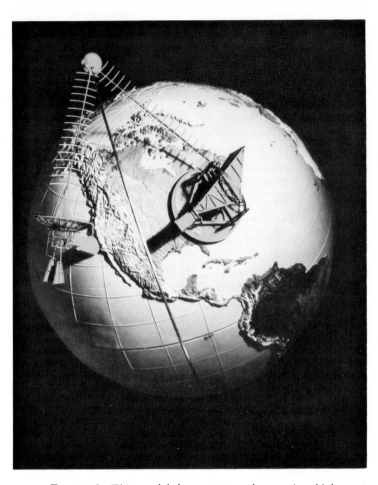

FIGURE 6 *This model demonstrates the way in which a satellite can beam data, television, or telephone communication to virtually any part of the world.*

basic research that led to the discoveries of quantum mechanics and the invention of the transistor and the solar cell, or without the electromagnetic discoveries of Maxwell and Hertz and the inventions of a host of radio pioneers. But they would have been just as impossible had these early ideas not been built upon and applied.

It is an incredibly long stride from the time Marco Polo could not convince his listeners of China's wonders to the time President Nixon entered Peking's Imperial Palace—and we watched it as it happened! Such global communication is an outgrowth of science and technology that is spreading rapidly over the face of the earth and that has been developed to its present state during only the last hundred of the thousands of years of man's history. That, alone, should be a cause for pride in intellectual progress. But with pride and knowledge we should also have humility. It may be that seeing basic human feelings such as fear, pity, anger, love, and frustration in so many guises will help us to understand better and to establish sensible relations with all the others with whom we share a small planet.

CHAPTER 16 GEORGE A. MILLER

Human Communication

> *Speech was made to open man to man, and not to hide him....*
> David Lloyd (1635-1692), in *The Statesman and Favourites of England Since the Reformation*

WHAT GOOD IS SCIENCE?

The answer depends on who gives it. To a scientist, enraptured by the exploration of ideas never before touched by human minds, unraveling mysteries beyond the imagination of his grandfather, the value of science is the value of beauty to the artist, strength to the athlete, faith to the clergy, knowledge to the historian. Science is valuable in itself as the ultimate expression of the human gift of rational thought.

But modern science is very expensive. Once scientists could themselves afford to indulge their strange weakness for understanding, but no more. Today the scientific enterprise requires large teams of men with long years of specialized training, an industrial society capable of providing the indispensible instrumentation, small cities to support the technical facilities for research and computation, vast libraries to make the published results accessible. Even a scientist who enjoys most deeply the intrinsic value of science—the full exercise of his intelligence, the race for new discoveries, the satisfaction that after him some puzzle need never again seem puzzling—even the purest scientist must ask himself how such investments of economic and human capital can be justified.

Modern science must be broadly supported by society. What good, then, is science to society?

Again, the answer depends on who gives it. To an industrialist, science is good business. The path from scientific discovery in the laboratory to technological innovations in industry has been traveled often enough to be almost predictable. Yet industry supports only a fraction of our scientific activity in the United States, and a very small fraction of that vital growing point we call "basic science." To an educator, science is a part of the culture that must be preserved and communicated to succeeding generations, another way to teach young people how to lean on their own minds. Yet the universities are scarcely able to maintain themselves; the vast resources required for modern science do not—indeed, could not—come from university endowments.

The major support for scientific research in the United States comes from the taxpayer. What good is science to him?

An economist might answer that scientific knowledge, like the air we breathe, is a common good, something from which everyone profits, but which no single person or group could afford to provide freely for everyone. Private enterprise would, therefore, underinvest in scientific research because its results could not be sequestered and appropriated by individuals. Even if scientific discoveries could be patented or copyrighted, they should not be, because the cost of spreading knowledge is negligible; the use of knowledge by one individual does not mean that less is available for others. Thus, the government is necessarily interested in scientific research, and the main burden of support falls on the taxpayer.

Economic arguments have a cold and abstract sound, however. A taxpayer wants to be reminded of more concrete values—of electric power, better medical treatment, faster transportation, higher agricultural yields, more powerful national defense, of almost everything that makes our lives today different from those of a century ago. These reminders are provided easily, and until very recently they were sufficient to justify public support of science.

Today, however, these scientifically based advances in modern technology are no longer viewed as unmixed blessings. The greater use of power is depleting our natural resources, better medical care has aggravated overpopulation, faster transportation is polluting our atmosphere, agricultural use of insecticides threatens our fellow creatures, atom bombs create a fear of world destruction. Today the public seems to be divided between wanting science to solve the problems that it has helped to create and wanting science to stop entirely until some of our more pressing social problems have been solved. A compromise between these views is frequently heard: our worst problems are largely social and economic in character, so stop supporting natural science and invest instead in social science. Such proposals, even when made with the best of intentions, misconstrue the nature and sources of scientific progress.

The optimal distribution of support among the diverse areas of science poses a difficult problem of national policy. No doubt each reader of this book will have his own priorities, but for most people, the position of the social scientist is likely to be higher today than ever before. No one knows better than the social scientist, however, how ill-prepared he is to shoulder all the burdens of a troubled society. Before social science can save society, society will have to strengthen social science. Merely needing a science is not sufficient to create one. If social science is to achieve a dignity equal to natural science, it must do so through the power of its ideas, not the importance of its problems. But that, of course, will take time—precious time.

The problems faced by social scientists are as difficult as they are urgent. Although promising beginnings have been made, many sceptics doubt that social science can ever supply the kind of technological control of events and processes that we have come to expect from natural science.

It is a matter of practical concern, therefore, to include some aspect of social science in this survey. Human communication seems particularly appropriate. Not only is communication the glue that holds society together; the scientific enterprise is itself a vast cooperative endeavor built on communication. No knowledge or wisdom could have accrued from the basic research reported in the preceding chapters without some form of human communication. It is communication that gives us our symbols for asking questions and formulating answers, for extending our logical powers and recording our observations and theories. No fact becomes a part of science until it has been communicated and accepted by other scientists. In effect, science itself is a form of human communication, without which civilization as we know it could not exist.

Scientific communication, of course, is merely one form of human communication—a highly specialized form that emphasizes generality and precision in the messages it encodes. In its cruder forms, human communication serves many other purposes connected with our continuing human efforts to get along with one another. These other forms of communication have themselves been the objects of scientific research. Indeed, of all the various natural phenomena that scientists have attempted to understand, the most fascinating, and probably the most baffling, is the process whereby one person becomes acquainted with others. Its fascination is that it comprises the bulk of what human life is about. And it is baffling not only because persons are so complicated, but because to study it is to exemplify it: the scientist must do just what laymen do, only more objectively and self-consciously. As the psychologist Liam Hudson has said, in asking that our passionate concern for other persons be knit together with our dispassionate concern for truth, facts, and intellectual analysis, we are asking for something at once profoundly important and profoundly complex.

Human communication, broadly conceived, encompasses all those processes whereby one person can become acquainted with another. In addition to participating in the kinds of social interactions every man participates in—observing, talking and listening and gesturing, reading and writing, cooperating and competing—the social scientist can use special tools—recording devices, physiological indicators, questionnaires, controlled environments—to collect his data. And in addition to the impressionistic inferences of the layman, the social scientist has statistical methods, logical or mathematical analysis, and all the resources of modern computer technology to help him decipher the meaning of his observations. These additional tools play such a conspicuous role in his work that, on the surface, the scientist's way of understanding other persons seems to bear little resemblance to the layman's. But these differences in means should not obscure the similarity in ends.

Just as a biologist can study organisms at levels ranging from the molecular to the ecological, so a social scientist can study human communciation at many different levels. Consider, for example, that uniquely human form of communication that we call speech. At the most physicalistic level of analysis, speech is an acoustic disturbance propagated as waves by displacements of the air molecules. It can be converted into electrical potentials by microphones and analyzed in complete detail by the instruments of modern communication science. Although this kind of research can tell us all we want to know about what speech *is*, it tells us almost nothing about what speech *does*. For social scientists, therefore, more abstract approaches are necessary.

Speech is the tangible manifestation of a system of social conventions that we call language. In order to describe a language it is necessary to describe its phonology, grammar, and lexicon. The phonology summarizes the distinctions among speech sounds that are significant for people who use the language. The grammar characterizes rules for combining sounds into well-formed words, phrases, and sentences. The lexicon lists all the basic words in the language along with some indication of their meanings. When such a description of a language is properly constructed, it includes all the information a student would need in order to learn the language.

Although linguistic descriptions of human language are considerably more abstract than are electro-acoustical descriptions of speech, they still fall short of our goal of explaining how speech helps one person become acquainted with another. Learning a language is merely the entrance fee for participating in the life of those who speak it. Analyzing what it is to know a language provides merely the foundation for studying what people use it for.

What are the functions of language? Many students of human communication have attempted to list them, and most lists include such functions as: to impart information, to elicit cooperation, to express emotions. A classification of functions is a slippery thing, however. The philosopher Ludwig Wittgenstein described some of the games we play with language: we give and obey orders, describe objects, report events, speculate about events, form and test hypotheses, make up stories, play-act, sing, guess riddles, tell jokes, translate, ask, thank, curse, greet, pray. Why should we believe that these diverse functions are finite in number, or that any short and generally useful catalogue of them could be constructed?

Consider an example. A husband returned home at the end of the day and was greeted at the door by his wife, who said, "I bought some light bulbs today." The husband understood exactly what she meant, but an outsider would not have. The outsider could have heard each word perfectly, could have parsed the sentence and defined every word of it, could have repeated or paraphrased it, and still would not have known that the wife was telling her husband, "If you want me to fix your dinner, you will have to replace that burned-out light in the kitchen before it gets dark." Such trivial facts as that a kitchen light had burned out, facts that could have no conceivable place in a grammar, lexicon, or encyclopedia, may play a critical role in understanding how a particular utterance is being used. On the surface, "I bought some light bulbs today" is a simple declarative sentence whose function is to impart information. To leave the description in that form, however, would be to omit the psychodynamics of that particular human interaction, without which the husband's reply, "Let's try that new restaurant," would have seemed a complete *non sequitur.* A science of human communication that had no place for the communicator's intentions would be like a car with an empty tank.

Another example. A small boy comes into the room where his parents are sitting and says, "I shot a million lions in the garden today." Such a sentence can be heard, parsed, interpreted, paraphrased, and even understood, but in few parts of the world would it be believed. What function does it serve? To treat it as a declarative sentence for imparting information would be to miss the point completely; to respond "You're a liar" would be brutally cruel.

Even the most casual observation of language in use provides convincing evidence that the functions of language are as diverse and subtle as the intentions, beliefs, hopes, and fears of the people who use it. Which means that all the resources of psychology, sociology, anthropology, linguistics, and philosophy must be brought to bear on the problem of understanding human communication. Yet every child comes to understand human communication before he starts to school. Why is it so difficult to do self-consciously and objectively what every child can do naturally and easily?

Most of the failures of communication that we see in our everyday lives are not failures of pronunciation or hearing, of grammar or lexicon. They are failures to understand adequately the intentions, beliefs, and desires of others. They are failures to communicate adequately our own intentions, beliefs, and desires. Recognition that human communication so frequently fails to satisfy our human needs is one of the sources of current interest in a special kind of education called sensitivity training.

The idea of sensitivity training was born almost accidentally in the summer of 1946 in the course of some studies by Kurt Lewin and his colleagues on how to gain a member's commitment to some course of action by a group. Its meteoric growth into a multimillion-dollar industry has affected psychotherapy and social psychology in profound ways. Schools, industries, and government have exploited the technique in several forms. Today, thousands of participants go every month to some center to experience the group methods. Kurt Back, who has recently recounted the history of this development, considers it to be as much a social movement as it is social science. The group experience seems to have met a felt need in our society, and the wide public interest and participation is in itself a noteworthy social phenomenon.

The essence of the technique is that a group of people come together for a limited period of time under the leadership of persons who focus their attention on the processes of social communication through explicit analysis, self-confrontation, immediacy of experience, and a variety of dramatic techniques calculated to break down the usual barriers of polite civility. Each person is given an immediate account of how he impresses the others, and is coached to concentrate on the group process itself and to confine his communications to a "here and now" orientation. These procedures usually induce strong emtional reactions, and when the emotional experience is made the center of the group's interest, it is usually called an encounter group. But the original purpose of sensitivity training was to teach people how to work and communicate more effectively in groups with other people.

The slogan of the movement is "here and now," a phrase that has attracted the attention of at least one linguist, Charles Fillmore. "Here" and "now" are what linguists call deictic words; their meaning can be understood only in the situation in which they are uttered. Like the personal pronouns "I" and "you," the meanings of "here" and "now" change, depending on who is using them where and when.

One of the members may say, "Life is rotten." The leader replies that this does not satisfy the here-and-now criterion. The member tries again: "I am telling you now that life is rotten." The leader again criticizes him, saying that it is not enough to describe his own utterance here-and-now; he is still being impersonal. So the member tries, "My life is rotten." Now the leader gets him on truthfulness: "Here you are drinking organic apple juice, surrounded by people who love you, and you tell us your life is rotten. How can you expect us to believe you?" Through a few more exchanges of this sort, the leader gradually helps the member create a sincere, here-and-now utterance, which is usually something like "I want you to feel sorry for me." At last he has produced a sentence between speaker and listener that is directly relevant to the moment of speech.

Although the procedures used are generally rationalized in the language of social science, there has been relatively little evidence for the scientific validity of these group sessions. The participants enjoy an intense emotional experience, become briefly but deeply engaged with the other members of the group, sometimes reach profound conclusions about themselves and the management of their lives, and usually leave with a conviction that it has been a useful, valuable experience. Many participate in much the same spirit they might attend the theater or take a winter cruise. It is not clear what a validation would look like in such cases. But it does seem clear that many people feel a desire to communicate more directly with other persons and need help in learning how to do it. Apparently there are implicit constraints on the ways it is acceptable to communicate in everyday affairs, constraints that protect us from too-easy intimacy with others. Sensitivity training is, at least in part, an attempt to restore freedom of speech or—since some advocates of sensitivity training believe that all words falsify experience—freedom of communication.

The goals of science are sometimes said to be the description, prediction, and control of natural phenomena. This characterization is probably justified by the historical development of the physical sciences. When these three goals are adopted uncritically by social scientists, however, they can lead to unfortunate consequences. To pinpoint the difficulty, consider the implications of saying that the goal of scientific studies of human communication is to control it. When this message reaches the public they cannot reconcile it with their traditional faith in the freedom of speech and the freedom of the press. Given a choice between a science of control and a philosophy of freedom, the layman finds little difficulty in reaching a decision.

To speak of social science as an instrument for social control is to do a disservice to both science and society. Even among physical scientists there are many who see the goal of science as understanding; prediction and control are merely techniques that can sometimes help to test our understanding. Leaving aside the question as to whether the science of human communication has really developed any new techniques for controlling human communication (techniques other than the traditional methods of oppression and censorship), the goal of the social scientist is understanding, not control; diagnosis, not prescription; description, not manipulation. He wants to do as scientist what every layman also wants to do—become better acquainted with other people.

Understanding other people, however, is a complex matter with implications of its own. For example, there is a theory that when two groups are in conflict, all you have to do to make them stop fighting is to bring them together so they can become acquainted. It is claimed that people who understand each other will stop fighting and rationality will then prevail. The social psychologist Muzafer Sherif has tested the idea by an experiment involving two groups of boys in a summer camp. First he contrived to bring the two groups into conflict (a maneuver that was soberingly easy to accomplish), and then arranged to bring the two groups together to share in various pleasurable activities so they could get to know one another. The result was disastrous; bringing them together simply provided an occasion for even more bitter conflict. Sherif's experiment substantiated what we see in the newspapers every day, namely, that merely bringing together people who are in conflict does not resolve that conflict.

Civilized society would be unthinkable without human communication, but communication is no panacea for our social ills. We can communicate aggression as readily as agreement, and there is certainly no guarantee that becoming acquainted with someone will cause us to like him any better. What the student of communication can contribute to the amelioration of social conflicts, however, is a better diagnosis of the real problems. All too often our attempts to intervene in personal or social problems fail as a consequence of defining the problem incorrectly. The most frequent mistake is to assume that the difficulty is somebody's fault, and that the guilty person must be identified and punished. The more upset one is about the difficulty, the greater is the temptation to blame it on a scapegoat. Much social science research can be cited to support the notion that defining the problem correctly is an indispensible step in solving it.

How can the real problem be defined? Here is where communication is most valuable. Assemble the people who are in a position to do something about the problem, give them a leader who knows how to avoid creating resistance, anxiety, or cognitive dissonance among the members—exactly the opposite strategy of the leader in an encounter group—and let them define the problem together. Once the problem is correctly defined, the solution will often be obvious, and it will usually not entail punishing or disciplining some scapegoat, or controlling someone's behavior with the latest techniques from the psychological laboratory. Some conflicts resist such a rational approach, but it is remarkable how often each individual participant will have a biased preconception of a shared problem, and how far communication can go toward its resolution.

But it must be communication of the right sort. If some members of the group are accused of causing the problem, they will resist the work of the group. If the anticipated solution threatens someone's security, he will become anxious and not contribute effectively. If the problem or its proposed solution seem contrary to someone's preconceptions, he will experience cognitive dissonance and refuse to accept it. An effective leader must be sensitive to these dangers and must have techniques that enable him to keep the group working harmoniously together. Indeed, if he does his job effectively, he will usually be so busy with the communication dynamics of the group that he will not have the time or the desire to impose his own preconceptions on their deliberations.

Understanding what is involved when one person communicates with others promises many social benefits in addition to the pleasures of intellectual insight into an enormously complex phenomenon. Such scientific investigations do not have as their goal the perfection of techniques for controlling the behavior of other people. They are intended to help us all do better one of the most important tasks we face in life—becoming acquainted with other persons and their ideas. And if they are successful, they can even help us become better acquainted with ourselves.

*   *   *

So we have come full circle, moving from outer space to what is sometimes called inner space. From there the circle starts again, for in man's inner space lie the seeds and the desire for new knowledge and understanding of himself and his world, the mysterious seeds that we do not understand intellectually at all, but that are the ultimate sources of all scientific progress.

CONCLUSION · GERARD PIEL

The Proper Study of Mankind

THE FIRST YIELD from work in science is science—that is, objective knowledge. Men can share such knowledge and act on it with confidence because they can repeat the work independently at other times and place. By this same procedure—by repetition of the work—science finds immense practical consequences; an industrial process is the original experiment repeated continuously on the grand scale. The utility of science has been celebrated in the other chapters of this book. The present chapter considers science as a kind of work that men do and that society ought to support for its own sake, as an end in itself.

Science thus purely defined is no less consequential in the lives of men. The self-recognition of the human species, across racial, national, cultural, and class boundaries, derives from scientific knowledge about man and the world around him and is shared ever more widely, with ascending moral force, among men all over the planet. The objectification of knowledge in science asserts the primacy of the individual man and makes irreversibly absurd the claim of any external authority upon his perception and judgment or upon his conscience. In his "flight from wonder"—the phrase is Albert Einstein's—man has come abroad into a universe that beggars in scale, invention, and splendor the paradises and infernos of his innocent imagination.

The choice of truth, the submission to the discipline of objective knowledge, is an ethical act. The truth is of itself a good, although it may be used for good or ill. Yet it is argued—and too often conceded—that science is value-free. To take that position is to abandon questions of value to unreason. As men comprehend that the natural locus of purpose in the universe is inside their own heads, they will learn to give these questions their best thinking. The immense utility of science compels such effort and lends pragmatic power to the golden rule.

The purity of science thus need not be argued against its utility. On the contrary, it is well to bear in mind that this kind of knowledge always implies control. The work of science is driven by questions that make a difference of one kind or another. Our times have seen great new technologies come from pure science. The improbable laser, for example, was implicit in the photoelectric effect. Yet, today as in the past, fundamental questions are posed by technologies at their frontiers. The physics of the solid state owes its revival to the crisis in telephone message-switching.

When it comes to the support of science, however, it is necessary to plead for science as an end in itself. While the outcome often motivates the patron, it does not always motivate the scientist. As has been shown again and again in this book, outcomes can only rarely be predicted. This country's first experience with the support of science from public funds, during the past 25 years, has accumulated ample evidence that funding by "mission-oriented" agencies can distort the distribution of resources and talent across the advancing front of knowledge. The availability of funds from military and paramilitary agencies brought many university scientists into relationships that compromised their independence.

I cite one example of such compromise: When the nuclear-weapons test-ban negotiations were getting under way in the early 1960s, most seismologists were confident that their instruments for identifying and locating earthquakes could do the same for underground tests of nuclear weapons, with satisfactory reliability down to very small energies of explosion. They soon found the Pentagon "Project Vela Uniform" pressing grants and contracts upon them to improve their instruments. Along with the grants and contracts, they accepted security-classification of their work; this effectively barred them from participation in public discussion of the feasibility of policing a ban on underground tests. In partial consequence, at least, underground tests were left out of the test-ban treaty. The arms race thereupon went into higher gear, with both sides testing weapons at a higher rate than before and without being hindered by protests against radioactive fallout.

Funding from mission-oriented sources has now proved to be undependable as well. Our country does not yet have an adequately chartered and funded source of public support for science. If the public is to support science wisely it must share more directly in the understanding of science.

There is deeper need for such sharing by the public. Science has subverted the traditional values of our culture and the comforts of its received absolutes, whether certified by religious or secular authorities. The resulting void is darkened by ignorance of science. Public anxiety cannot much longer be appeased by the distractions of affluence and gadgetry. Issues of value and purpose have gathering urgency in American culture. The understanding necessary to secure the ground to resolve these issues is not widely shared, even among the large minority of the citizenry that has been exposed to higher education.

There can be no turning back to the Lilliputian presumption of man's claim that he stands at the center of the cosmos. A succession of Copernican revolutions has removed the sun as well as the earth from the center and placed them in an ordinary lane of stars in a common type of spiral galaxy located nowhere in particular in an ever more vast and ancient universe. As man proceeds with the recognition of his identity, however, he may take what cheer he can from the knowledge that the universe has no center and no boundaries, and so affords no other more privileged outlook on its past and present.

Darwinian evolution similarly erased the fantasy of special creation. Biochemical charting of the family tree shows man to be the natural child of this planet and a cousin to every living thing on it. The sequence of amino-acid subunits in the beta-hemoglobin molecule in human blood differs at only one site from that sequence in the gorilla. Difference at only 10 sites suggests that the pig is more closely related to man than is the horse, which differs at 26 sites. Comparison of the amino-acid chains in cytochrome-C, the molecule that mediates combustion in the cell, distinguishes man and his primate cousins from the other mammals by a half-dozen amino-acid substitutions, on the average, in the chain of 104; the higher vertebrates from the fishes by 19 substitutions, and the vertebrates from the insects by 27 substitutions. Essentially the same ancient cytochrome-C presides at the identical reaction in all living cells; its amino-acid sequence differs, at the extreme, by only 40 per cent between man and the humble bread mold, *Neurospora.*

Carbon compounds identified in meteorites and, by radio telescopes, in outer space indicate that life itself is not so special a creation. Simple compounds form spontaneously in the statistically rare, yet sufficiently frequent, encounters of atoms wandering over eons of time in the interstellar vacuum. In the laboratory, quite uncomplicated experimental procedures produce such higher-order compounds as amino acids and the nucleotide subunits of DNA, the molecule of heredity. These compounds are, it appears, the inevitable outcome of the jostling together of carbon, hydrogen, oxygen, and nitrogen under suitable, but not stringent, conditions. Life, in the metaphor of the French physicist Pierre Auger, is a ripple, a standing wave, in the flood of energy and matter that streams downhill into the cold emptiness of space from the interior of stars.

The apparent universality and inevitability of life press the further hypothesis that intelligence may be an equally universal and inevitable culmination of evolution wherever life has ignited in this and other galaxies. If it can lead to man, then why not beyond, to higher intelligence? We may find such comfort or anxiety, as we choose, in the growing evidence that we are not alone.

As a matter of degree, rather than of kind, intelligence is not even a local, terrestrial monopoly of man. The chimpanzee has demonstrated a capability for symbolic communication. Through graphic symbols it can arrange with its hands, instead of the spoken word, this animal can express itself—to humans, if not to other chimps— in subject-predicate sentences. Whether a computer can be made to think is a question that remains open only in semantics. For some persons, the notion remains abhorrent; it makes the human brain, they say, nothing but a "meat machine." This is an obscenity, but only if one thinks meat and machines obscene.

The hard, Puritan injunction that follows from man's objective knowledge of man is that he takes his identity —his dignity, his divinity—from nothing and from nowhere but himself. Each human consciousness is a unique, never to be re-enacted, episode. The example of the best and the worst shows men to be potentially as richly unique and individuated as if each were himself a species. The individual realizes his potential, however, only by living in society with other men.

How identity is won in interdependence is illustrated by work in science. No scientific work stands alone; it derives its significance from its relation to the past, from which it may be a departure, and to the future, which it must urge forward. Scientists address their work to one another across national boundaries and across generation gaps. The liberty to realize individual potential is asserted not in isolation but in context. We who are alive today are witnessing the advent of the first truly world culture; through the common language of reason, all men can share in the expansion of the human consciousness.

Talk of a "knowledge explosion" misses the true nature of this turning point in history. Such talk proceeds from the mistaken notion that science is mere fact-piling. Immense volumes of data, it is true, are involved. But no one needs to know it all. What is necessary is to understand the ideas that organize the facts, connect them to one another and give them meaning. No fact gets established without an idea; sometimes wrong, as well as right, ideas lead to new facts. The significance of a discovery is most often measured not by how many questions it answers but how many it asks. Science makes progress by revolutions in ideas that reorganize the facts. In this respect, the history of science finds many parallels in the history of the arts.

This much having been said, it is necessary to emphasize the empirical, fact-establishing aspect of the work. An idea carries the day not by the originality, attractiveness, or persuasiveness of itself or its author, but by experiment or observation. On the facts and what they show, no scientist can accept any authority but his own judgment. Yet every scientist must submit his work to verification by the sovereign authority of each of those of his brothers who finds his work interesting enough to make him want to repeat it.

There are many troubling questions, it is true, about this method of getting at the truth. The logic that ties the evidence to the inference has big gaps and holes in it, compared to the tidy operations of deductive logic and school philosophy. A different and deeper uncertainty besets the experiment, in that the observation not only perturbs the event but does so to an uncertain degree.

Epistemological doubts about the ground of scientific knowledge are settled, however, by the observed fact that it works. It has extended its reach into larger and larger realms of human experience over the past 400 years. In fact, the pragmatic, empirical way of knowing is the only way men really know anything. The claims for other ways—by "intuition," "feeling," "insight," "a priori knowledge"—delimit those realms of experience we do not yet know and understand so well. There and by these other ways it is still possible to know certainties and to cherish absolutes. Science requires a stoic will to live with uncertainty.

"The aim of a scientific exposition," said Richard von Mises, "can never be other than an intellectual one, that is to say, to offer information, enlightenment, elucidation of relationships, no matter what feelings of pleasure or displeasure result from these." To this it should be added that von Mises found great pleasure in his work in aerodynamics and wrote his philosophical treatises with precision and felicity. He would not deny the esthetics that lure the scientist to his work and rewards the layman in his vicarious participation. The proposition that order underlies the prodigality of nature and that uniformities govern its exhuberant forces is a trusting, human one. But it is sustained at every turn. Beneath the ever-changing appearances of things, scientific inquiry discloses proportion, harmony, symmetry, economy, and elegance of design, exquisite elaboration upon simple themes; nature matches every paradigm of beauty that the mind can hold up to it. This is no surprise, of course, because the mind also is a natural invention. Scientific explanation is not the reduction of higher-order phenomena to lower-order terms. In the words of George Wald, physics and chemistry, in approaching the explanation of life, "will have grown up to biology." Rational understanding transcends wonder, because it is a mode of perception that sees deeply and more clearly.

There is thus—again the phrase in Einstein's—a "passion to understand." Passion is indeed required to motivate the detached objectivity that puts the hypothesis to the test of evidence. A false and fashionable dichotomy nowadays separates "affective" and "cognitive" knowledge. This schizophrenia works its worst mischief when questions of value are at issue. It has a dishonorable origin in the truce drawn between the faculties of arts and of sciences that closed the Darwinian controversy a century ago. Science, held to be value-free, was reserved to scientists. Value was correspondingly staked out for the humanists, as lying beyond the reach of reason in the province of feeling and emotion, which it was the function of arts and letters to refine and elevate.

For the most part, this division of labor continues to prevail in the faculties. Its premises are widely accepted in the popular culture. Lyndon Johnson, in affixing his presidential signature to the legislation establishing the National Foundation for the Arts and Humanities, was moved to declare: "We need science to make our goods. We need the humanities to give us our values."

Such dualism too often sanctions the employment of bad means for the attainment of declared good ends. If the ugly lessons of history are rightly read, it can be seen that the means always, ultimately, comprehends ends. The evil act is the terminus of good intentions.

The pragmatic philosophy of science applied to questions of value declares that purpose must find validation in the same tests that reason applies to truth. As men's understanding grows and their technological competence increases, men's values change. "Such concepts as justice, humanity, and the full life," says the mathematician J. Bronowski, "have not remained fixed in the last hundred years, whatever churchmen and philosophers may pretend. . . . A civilization cannot hold its activities apart or put on science like a suit of clothes—a workday suit which is not good enough for Sundays."

Consider the notion of "race" and the role it has played in history. Those who held citizenship in the democracy of Athens could speak of their countrymen outside the city gates as "andropoda," that is, human-footed animals. So, wherever the minority of the powerful held the powerless majority in subjection, whether as helots, serfs, or chattel slaves—and that was everywhere in agricultural civilization—they could make of their subjects what the psychologist Erik Erikson calls a "pseudospecies." The proposition was most often sustained by distinct disparity in biological development as between masters and subjects, when it was not reinforced by color. That the disparity was due to inequity can be seen today in the contrast in height and weight between the prerevolution Russian peasantry and their postrevolution urban grandchildren. The evidence amassed by genetics, anthropology, psychology, and human biology has overwhelmingly crushed the mythology of race. Such a concept can no longer hold any standing in the value system of a civilized society or supply rationalization for social policy. On the contrary, objective knowledge affirms the value of equality before the law.

The ethic of objective knowledge sets noble aims for the conduct of the individual and for the organization of society. In the words of the biochemist Jacques Monod:

In the ethic of knowledge the single goal, the supreme value, the "sovereign good" is...objective knowledge itself.... There must be no hiding that this is a severe and constraining ethic which, though respecting man as the sustainer of knowledge, defines a value superior to man. It is a conquering ethic...since its core is a will to power: but power only in the [realm of ideas and knowledge]. An ethic which will therefore teach scorn for violence and temporal domination. An ethic of personal and political liberty; for to contest, to criticize, to question constantly are not only rights therein, but a duty. A social ethic, for objective knowledge cannot be established as such elsewhere than within a community which recognized its norms.

These are the values of the emergent world culture. It is no coincidence that they embrace the ideals of the open society: liberty, justice, and brotherhood. For it is the scientific-industrial revolution that has made possible the approach to this social order by the fortunate peoples who live in the industrial, or "developed," countries. The enormous power that applied science has placed in man's hands now makes urgent the ordering and perfecting of human society. People are acting, however, as if they are used to living with the threat of the thermonuclear apocalypse. At the same time, they are too little informed of the positive measures at their command by which they might terminate the age-old rule of want and inequity. But what men need most to learn is the liberating knowledge about themselves already established by the work of science.

SUGGESTED ADDITIONAL READING

CHAPTER 1 *Hoyle*

Jean A. Charon. Cosmology. McGraw-Hill, New York, 1970.

George Gamow. Creation of the Universe. Viking, New York, 1952.

Paul W. Hodge. Physics and Astronomy of Galaxies and Cosmology. McGraw-Hill Science Series, New York, 1966.

Fred Hoyle. Galaxies, Nuclei and Quasars. Harper, New York, 1965.

CHAPTER 2 *Langbein*

L.B. Leopold and K.S. Davis. Water. Life Science Library, Time, Inc., 1966.

R.J. Chorley (Editor). Introduction to Physical Hydrology. Harper and Row (Barnes and Noble Import Division): Methuen, 1969.

CHAPTER 3 *Battan*

L.J. Battan. Harvesting the Clouds. Doubleday and Co., Garden City, N.Y., 1969.

L.J. Battan. Weather. Prentice-Hall, Englewood Cliffs, N.J., 1974.

G.E. Dunn and B.I. Miller. Atlantic Hurricanes. Louisiana Univ. Press, Baton Rouge, La., 1960.

Inadvertent Climate Modification: Report of the Study of Man's Impact on Climate. MIT Press, Cambridge, Mass., 1971.

M. Riehl. Introduction to the Atmosphere. McGraw-Hill Book Co., New York, 1965.

The Atmospheric Sciences and Man's Needs. National Academy of Sciences, Washington, D.C., 1971.

CHAPTER 4 *Jahns*

James Gilluly, A.C. Waters, and A.O. Woodford. Principles of Geology. (Third edition). W. H. Freeman and Company, San Francisco, 1968.

CHAPTER 4 *Jahns (continued)*

K.F. Mather. The Earth Beneath Us. Random House, New York, 1969.

W.C. Putnam. Geology (Second edition, revised by Ann B. Bassett). Oxford University Press, New York and London, 1971.

J.S. Shelton. Geology. W.H. Freeman and Company, San Francisco, 1966.

H. Takeuchi, S. Uyeda, and H. Kanamori. Debate About the Earth. (Revised edition). Freeman, Cooper and Company San Francisco, 1971.

CHAPTER 5 *Eccles*

John C. Eccles. Facing Reality: Philosophical Adventures by a Brain Scientist. Springer-Verlag, New York, 1970.

S. Ochs. Elements of Neurophysiology. John Wiley & Sons, New York, 1965.

L.A. Stevens. Explorers of the Brain. Alfred A. Knopf, New York, 1971.

Samuel Bogoch (Editor). The Future of the Brain Sciences. Plenum Press, New York, 1969.

John C. Eccles. The Understanding of the Brain. McGraw-Hill, New York, 1973.

CHAPTER 6 *Segal*

S. Fred Singer (Editor). Is There an Optimum Level of Population? McGraw-Hill Book Company, N. Y., 1971.

Egon Diczfalusy and Ulf Borell (Editors). Nobel Symposium 15: Control of Human Fertility. Almqvist and Wiksell, Stockholm, Wiley Interscience Division, John Wiley & Sons, New York, London, Sydney, 1971.

Clyde V. Kiser (Editor). Forty Years of Research in Human Fertility. Milbank Memorial Fund, Volume XLIX, Number 4, Part 2, October, 1971.

Bernard Berelson, Richmond K. Anderson, Oscar Harkavy, John Maier, W. Parker Mauldin, and Sheldon J. Segal (Editors). Family Planning and Population Programs: A Review of World Developments. The University of Chicago Press, 1966.

S.J. Behrman, Leslie Corsa, Jr., and Ronald Freedman (Editors). Fertility and Family Planning: A World View. The University of Michigan Press, Ann Arbor, 1969.

CHAPTER 7 *Mayer*

Stanley Davidson and R. Passmore. Human Nutrition and Dietetics (4th ed). Williams and Wilkins, Baltimore, 1969.

Ronald M. Deutsch. The Family Guide to Better Food and Better Health. Meredith, Des Moines, 1971.

J. Mayer. Overweight: Causes, Cost and Control. Prentice-Hall, Englewood Cliffs, N.J., 1968.

J. Mayer. Human Nutrition. Charles C Thomas, Springfield, Ill., 1972 (in press).

J. Mayer. U.S. Nutrition Policies in the Seventies. W.H. Freeman, San Francisco, 1972 (in press).

CHAPTER 8 *Nossal*

F. M. Burnet. Self and Not Self, Book I. Cellular Immunology. Cambridge University Press, London and New York, 1969.

P.B. Medawar. The Uniqueness of the Individual. London, Methuen, 1957.

G.J.V. Nossal. Antibodies and Immunity. Basic Books, N. Y., 1969.

G.J.V. Nossal. "How Cells Make Antibodies." *Scientific American* 211: 106, 1964.

I.M. Roitt. Essential Immunology. Oxford, Blackwell Scientific Publications, 1971.

CHAPTER 9 *Hogness*

A.M. Srb, R.D. Owen, and R.S. Edgar (Editors). Facets of Genetics. W. H. Freeman and Company, San Francisco, 1970. (An excellent collection on all aspects of genetics. Taken from articles in *Scientific American* that are quite suitable for the layman. Paperback.)

W.S. Laughlin and R.H. Osborn. Human Variations and Origins. W. H. Freeman and Company, San Francisco, 1967. (Another interesting collection of papers for the layman from *Scientific American*, this time concentrating on human heredity. Paperback.)

James D. Watson. Molecular Biology of the Gene. (Second Edition.) W.A. Benjamin, Inc., New York, 1970. (A lucid description of this subject in simple language. Paperback.)

Chapter 10 *Mark*

F.W. Billmeyer, Jr. Textbook of Polymer Science. John Wiley & Sons, New York, 1972.

A.V. Tobolsky and H. Mark. Polymers in Material Science. John Wiley & Sons, New York, 1972.

Giant Molecules. Life Science Library, Time-Life Books, New York, 1966.

H.F. Mark. "Polymeric Materials." In *Scientific American*, Vol. 217, No. 3 (Sept. 1967).

H.F. Mark. Polymers—Past, Present, Future. Proceedings of the R. A. Welch Foundation Conference on Chemical Research No. 10, pp. 19–55; published in 1967.

Chapter 11 *Schawlow*

Vasco Ronchi. The Nature of Light. Heinemann, London, 1970.

Humberto Fernández-Morán. The World of Inner Space. In *Science Year* 1968, pp. 216–227. Field, Chicago, 1968.

Jean-Claude Pecker. Space Observatories (Translated from the French by J.R. Lesh). Springer-Verlag, New York, 1970.

Scientific American Readings in Lasers and Light. (With Introductions by A. L. Schawlow). W. H. Freeman, San Francisco, 1969.

L. Goldman and R.J. Rockwell, Jr. Lasers in Medicine. Gordon and Breach, New York, 1971.

Chapter 12 *Kac*

Mathematics in the Modern World. Readings from *Scientific American*. W.H. Freeman and Company, San Francisco, 1968.

Computers. S.M. Ulam. pp. 337–346.

The Uses of Computers in Science. A.G. Oettinger. pp. 361–369.

Man Viewed as a Machine. J.G. Kemeny. pp. 386–393.

Chapter 13 *McKelvey*

H.J. Barnett and Chandler Morse. Society and Growth. Johns Hopkins Press, Baltimore, 1963.

Donald A. Brobst and Walden P. Pratt (Editors). United States Mineral Resources: U.S. Geol. Survey Prof. Paper 820; Summary in U.S. Geol. Survey Circular 682, 1973.

Harrison Brown. The Challenge of Man's Future. Viking Press, New York, 1954.

Hans H. Landsberg and Sam H. Schurr, Energy in the United States: Sources, Uses, and Policy Issues. Random House, New York, 1968 (paperback).

Donella H. Meadows, Dennis L. Meadows, Jorgen Randers, and William W. Behrens, III. The Limits to Growth. Universe Books, New York, 1972 (paperback).

National Academy of Sciences—National Research Council. *Resources and Man.* W.H. Freeman, San Francisco, 1969 (paperback).

Brian J. Skinner. Earth Resources. Prentice-Hall, New Jersey, 1969 (paperback).

CHAPTER 14   *Starr*

United States Atomic Energy Commission. Understanding the Atom Series. USAEC, P.O. Box 62, Oak Ridge, Tenessee 37830.

Samuel Glasstone. Sourcebook on Atomic Energy. Van Nostrand Publishing Co., 1967.

Ralph E. Lapp. Atoms and People. Harper Bros., 1956.

William R. Anderson and Vernon Pizer.   The Useful Atom. World Publishing Co., 1966.

David O. Woodbury. Atoms for Peace. Dodd, Mead & Co., 1965.

Scientific American. "Energy and Power" Issue. September 1971.

CHAPTER 15   *Pierce*

Arthur C. Clarke. Voice Across the Sea. Harper and Bros., New York, 1958.

Arthur C. Clarke. Voices from the Sky. Harper and Row, New York, 1965.

J.R. Pierce. The Beginnings on Satellite Communications. San Francisco Press, San Francisco, California, 1968.

The Encyclopedia Brittanica. Various articles, including "Telegraph," "Telephone."

CHAPTER 16 *Miller*

Kurt W. Back. Beyond Words: The Story of Sensitivity Training and Encounter-Groups. Russell Sage Foundation, New York, 1972.

Liam Hudson. The Cult of the Fact. Jonathan Cape, London, 1972.

George A. Miller. The Psychology of Communication: Seven Essays. Basic Books, New York, 1967.

John R. Pierce. Communication. In *Scientific American*, September, 1972, Vol. 227, No. 3, pp. 31–41.

M. Sherif, O.J. Harvey, B.J. White, W.R. Hood, and C.W. Sherif. Intergroup Conflict and Cooperation: The Robbers Cave Experiment. Institute of Group Relations, University of Oklahoma, Norman, Oklahoma, 1961.

Jacobo A. Varela. Psychological Solutions to Social Problems: An Introduction to Social Technology. Academic Press, New York, 1971.

BIOGRAPHIES

In order of author's appearance in book

DR. ISAAC ASIMOV was born in 1920 in the Soviet Union, was brought to the United States in 1923, and has been an American citizen since 1928. He joined the Biochemistry Department of the Boston University School of Medicine in 1949 and now holds the rank of Associate Professor. In 1938, he began a career as a writer, first in science fiction, then in a wide-ranging nonfiction field that includes popularized science, history, literature, humor, and so on. He writes for all ages, from textbooks for medical students to picture books for seven-year-olds. At the present moment he has published 122 books and has 12 in press.

PROFESSOR SIR FRED HOYLE was Plumian Professor of Astronomy and Experimental Philosophy in the University of Cambridge, England, and Director of the University's Institute of Theoretical Astronomy from 1958 to 1972. In 1973 he joined the faculty of the University of Manchester. He also is a staff member of the Mt. Wilson and Palomar Observatories in California, and Visiting Professor at the California Institute of Technology. He is well known as a lecturer, and has written widely, including books on cosmology, sociology, and science fiction. He is a Fellow and Vice President of the Royal Society and served on the Science Research Council (of Great Britain). In 1968 he was awarded the Kalinga Prize of the United Nations, and received his knighthood in 1972.

DR. WALTER B. LANGBEIN joined the U. S. Geological Survey in 1935 as a hydraulic engineer-hydrologist, later moving into hydrological research in the U.S.G.S. He received a Doctor of Science (*honora causa*) degree from the University of Florida, was elected to the National Academy of Sciences, and has received the following honors: Distinguished Service Award from the Department of the Interior; the Bowie Medal and the Robert E. Horton Award from the American Geophysical Union; the Stevens Award from the American Society of Civil Engineers; and the Gano Dunn Award from Cooper Union.

DR. LOUIS J. BATTAN, Professor of Atmospheric Sciences and Associate Director of the Institute of Atmospheric Physics at the University of Arizona, was president of the American Meteorological Society in 1966-67, and received both the Meisinger and the Brooks awards of that Society. He has served on many advisory committees of various government agencies, is a member of the Committee on Atmospheric Sciences of the National Academy of Sciences, and is the author of 10 books, which range from college texts to nontechnical paperbacks on atmospheric sciences, and of many articles in scientific journals.

DR. RICHARD H. JAHNS is Professor of Geology and Dean of the School of Earth Sciences at Stanford University. His professional career has included work with the U. S. Geological Survey in many parts of this country, as well as 27 years of teaching and research at Caltech, Pennsylvania State University, and Stanford. With principal interests in the fields of petrology and economic, engineering, glacial, and structural geology, he has long been fascinated by the complex problems of man's peaceful coexistence with his physical environment. This has led him into studies of such dynamic earth phenomena as floods, landslides, and earthquakes.

PROFESSOR SIR JOHN C. ECCLES was born in Melbourne, Australia, and received his medical degree from the University of Melbourne. He went to Oxford University as a Rhodes Scholar, received his D. Phil. from that University, and remained associated there for several years until he returned to Australia. Since 1966 he has been in the United States, and now is Distinguished Professor of Physiology and Biophysics, State University of New York in Buffalo. He received the Nobel Prize for Medicine in 1963, is the recipient of many other awards and honors, and is the author of six books, in addition to many articles in scholarly journals.

DR. SHELDON J. SEGAL is an embryologist and reproductive physiologist who is Vice President and Director of the Biomedical Division of the Population Council. He has been at The Rockefeller University as an Affiliate and as an officer of the Population Council since 1956. He serves, also, as a member of the World Health Organization Advisory Group on Human Reproduction, the scientific advisory panel of Harvard University's Laboratory of Reproduction Research, and as a member of the U. S. Food and Drug Administration's Advisory Committee on Obstetrics and Gynecology. Presently, he is a special consultant on research to the Assistant Secretary for Health and Scientific Affairs, Department of Health, Education, and Welfare. Dr. Segal is President of the International Society for Reproduction Research.

DR. JEAN MAYER is Professor of Nutrition, Lecturer on the History of Public Health, and Member of the Center of Population Studies at Harvard University. He has worked in technical assistance projects for various United Nations agencies, served as Special Consultant to the President of the United States, Chairman of the White House Conference on Food, Nutrition, and Health, and is a member of the President's Consumer Advisory Council and Chairman of its Health Committee. He writes a column on nutrition, syndicated in more than 80 of the country's largest newspapers, and is a regular contributor to *Family Health* magazine.

DR. G. J. V. NOSSAL is Director of the Walter and Eliza Hall Institute of Medical Research, Melbourne, Australia, and Professor of Medical Biology in the University of Melbourne. He is widely known as a theorist and international lecturer in antibody formation and immunological tolerance. He is also concerned about communication between scientist and layman, and two of his three books, *Antibodies and Immunities* and *Medical Science and Human Goals* are aimed in this direction. Among his numerous distinctions are the Emil Von Behring Prize, the Phi Beta Kappa Science Award, and the Fellowship of the Australian Academy of Science.

DR. DAVID S. HOGNESS is a professor in the Department of Biochemistry, Stanford University. His research during the late 50's and early 60's resulted in the mapping of genes in chemical units, and was instrumental in demonstrating the relationship between genetic and chemical maps. He received the Newcomb Cleveland Prize in 1966 for this work. In 1969, with the aid of a Guggenheim Fellowship, he shifted his research interests to the organization and expression of genes in animal chromosomes.

DR. HERMAN F. MARK was born in Vienna, Austria, and received his doctorate from the University of Vienna. In 1938, after several years of research in Austria, he became Research Manager of the Canadian International Paper Company in Hawkesbury, Ontario, Canada. Since 1940, Dr. Mark has been associated with the Polytechic Institute of Brooklyn as Professor, Director of the Polymer Research Institute, Dean of the Faculty, and Dean Emeritus. He has published over 500 articles in many American and foreign journals, and has written about 15 books on various topics related to polymer chemistry. Dr. Mark is Editor or Associate Editor of three scientific journals, and is associated with many others.

DR. ARTHUR L. SCHAWLOW was a research physicist at Bell Telephone Laboratories until 1961, when he became Professor of Physics at Stanford University. With C. H. Townes, Dr. Schawlow is coauthor of a book, *Microwave Spectroscopy*, and of the first paper describing optical masers, which are now called lasers. For this latter work, Schawlow and Townes were awarded the Stuart Ballantine Medal by the Franklin Institute (1962), and the Thomas Young Medal and Prize of the Physical Society and the Institute of Physics (1963). He holds three honorary doctorates and is a member of several professional and scientific societies, including the National Academy of Sciences. He wrote the introductions for *Scientific American Readings on Lasers and Light,* and three of the articles in that collection. He has appeared on television for American, Canadian, and British networks.

Born and educated in Poland, DR. MARK KAC joined the faculty of Cornell University in 1939, and became a Professor at The Rockefeller University in 1961. A winner of the Chauvenet Prize of the Mathematical Association of America in 1950 and again in 1968, he is the author of numerous articles and several books, the latest of which (written jointly with S. M. Ulam) is *Mathematics and Logic*, a popular exposition of trends and themes of contemporary mathematics. Professor Kac is a member of the National Academy of Sciences, the American Philosophical Society, and the American Academy of Arts and Sciences.

In 1972, DR. V. E. McKELVEY became the ninth Director of the U. S. Geological Survey in its 93-year history. He joined the Survey on completion of his graduate work at the University of Wisconsin in 1940, and became Chief Geologist in 1971. He was Chief of the Survey's radioactive minerals investigations from 1950 to 1953, has served on advisory committees for this country and the U. N., and has acted as consultant or adviser in the Philippines, Jordan, and Saudi Arabia. He is a member of several scientific and professional societies, and in addition to other honors and awards, received the Department of the Interior Distinguished Service Award in 1963.

DR. CHAUNCEY STARR, President, Electric Power Research Institute, Los Angeles, was Dean of the UCLA School of Engineering and Applied Science from 1967 to 1973, after a 20-year career as an executive with North American Rockwell Company. He pioneered in the development of nuclear propulsion for rockets and ramjets, in miniaturizing nuclear reactors for space, and in developing atomic-power electricity plants. Dr. Starr is Vice President of the National Academy of Engineering, and member of the Environmental Studies Board of the National Academy of Sciences-National Academy of Engineering. He is a founder and past president of the American Nuclear Society, and a past vice president of the Atomic Industrial Forum. He is also a member of various governmental committees, and is a Foreign Member of the Royal Swedish Academy of Engineering Sciences.

DR. JOHN R. PIERCE is Professor of Engineering at the California Institute of Technology. He joined Bell Telephone Laboratories in 1936, and when he left for Caltech in 1971 he was Executive Director, Research-Communications Sciences Division. His research on the traveling-wave tube led to its application in communication satellites, in which he became interested in 1954. The Bell Laboratories work in connection with the Echo satellite (1960) and the TELSTAR ® satellite (1962) are based on his original suggestions. Dr. Pierce is a member of the National Academy of Sciences and the National Academy of Engineering, and was awarded the National Medal of Science in 1963. He is the author of 13 books, many of which have been translated into several languages. He is also a member of the Science Fiction Writers of America, and has published a number of stories under his own and other names.

DR. GEORGE A. MILLER was graduated from the University of Alabama in 1940 and received his Ph.D. from Harvard University in 1946 for research on military voice communications. Until 1968 he worked at Harvard and M. I. T. on the psychology of communication and the cognitive implications of language. He was appointed Professor at The Rockefeller University in 1968, where he has devoted his attention primarily to semantics and to experimental studies of meaning. Dr. Miller is a member of the National Academy of Sciences, and served as President of the American Psychological Association in 1968-69.

After serving as science editor of *Life* magazine from 1939 until 1945, GERARD PIEL was assistant to the president of the Henry Kaiser Company and associated companies until 1946. In 1948, he and two colleagues launched the new *Scientific American*. Mr. Piel holds several honorary degrees, and is a member or fellow of the American Philosophical Society, the American Academy of Arts and Sciences, and the Institute of Medicine of the National Academy of Sciences, and serves as a Trustee of Radcliffe College, Phillips Academy, The American Museum of Natural History, the Mayo Foundation, and as a member of various visiting committees of the Overseers of Harvard College. He was awarded the UNESCO Kalinga Prize in 1962, the George K. Polk Award in 1964, the Branford Washburn Award in 1966, and the Arches of Sciences Award in 1969.

ILLUSTRATION CREDITS

INTRODUCTION
Quotation. T. Dobzhansky. *The Biology of Ultimate Concern*, New American Library, New York, 1967, p. 9.

CHAPTER 1 *Hoyle*
Table 1. From the *Journal of the Astronomical Society of the Pacific*, November-December, 1972.
Figure 1. Hale Observatory.
Figure 2. Mt. Wilson and Palomar Observatories.

CHAPTER 2 *Langbein*
Quotation. George Gaylord Simpson. *The Meaning of Evolution*. (Revised edition) Bantam, N.Y., 1971, p. 265.
Figure 4. Donald G. Polovitch/Rapid City, S.D. *Journal*.
Figure 5. Wm. S. Russell/Bureau of Reclamation.

CHAPTER 3 *Battan*
Figure 1. NASA.
Figure 2. After Quiroz, *Bull. Amer. Meteor. Soc.*, Feb., 1972, p. 123.
Figure 3. NOAA/National Weather Service.
Figures 4,5. (ESSA) Environmental Science Services Administration

CHAPTER 4 *Jahns*
Figure 1. J.R. Balsley/United States Geological Survey.
Figures 4,5. Geotronics.

CHAPTER 5 *Eccles*
Quotation. C.H. Waddington. *The Nature of Life*. Atheneum, N.Y., 1961, p. 124.

CHAPTER 7 *Mayer*
Figure 2. Courtesy Academic Press.
Figure 3. Courtesy AID.

CHAPTER 8 *Nossal*
Figure 5A. Courtesy Dr. D.R. Davies, Laboratory of Molecular Biology, National Institute of Arthritis and Metabolic Diseases, NIH.
Figure 5B. Courtesy Dr. G.M. Edelman, The Rockefeller University.

CHAPTER 9 *Hogness*
Quotation. George Wald. In *Evolutionary Biochemistry*, Vol. 111., (A.I. Oparin, Editor). Pergamon Press, Ltd. 1963, p. 12.

CHAPTER 10 *Mark*
Quotation. J. Robert Oppenheimer. "The Need for New Knowledge." In *Symposium on Basic Research*, AAAS Pub. #56, 1959.
Collage by Gerald McConnell.

CHAPTER 11 *Schawlow*
Figures 1,2. Burndy Library.
Figure 3. Courtesy Grumman Aerospace Corporation.
Figure 4. Courtesy Dr. Humberto Fernández-Morán, University of Chicago.
Figure 5. Courtesy Dr. Erwin Müller, Pennsylvania State University.
Figure 6. Courtesy Dr. Edwin R. Lewis, University of California at Berkeley.

CHAPTER 12 *Kac*
Quotation. Joseph Weizenbaum. In *Science*, May 12, 1972.
Figure 1. Burndy Library.

Cartoon by W. Miller: © 1965, *The New Yorker*, Inc.

CHAPTER 13   *McKelvey*

Figure 1.   Data from United Nations 1967 *Statistical Yearbook*.

Figure 2.   Data on energy production from M.K. Hubbert, *Scientific American*, September, 1971; data on population from Donald J. Bogue, *Principals of Demography*, John Wiley & Sons, New York. 1969.

Figure 3.   U.S. Bureau of Mines.

Figure 4.   U.S. Geological Survey.

CHAPTER 14   *Starr*

Quotation. J. Bronowski. In *The American Scholar,* Spring 1972, p. 207.

Figure 4.   From *Nuclear Power Plants*, by Robert L. Loftress. © 1962 by Litton Educational Publishers, Inc. Reprinted by permission of Van Nostrand Reinhold Company.

Figure 6.   AEC

CHAPTER 15   *Pierce*

Quotation. James B. Conant. *Science and Common Sense.* Yale University Press, New Haven, 1951, p. 6.

Figures 1,4A.   Burndy Library.

Figures 2,3,4B,5,6.   Bell Telephone Laboratories.

Index

Note: Numbers in italics refer to pages on which tables and figures appear.